From Newton to Einstein

Ask the physicist about mechanics and relativity

From Newton to Einstein

Ask the physicist about mechanics and relativity

F Todd Baker

Morgan & Claypool Publishers

Rights & Permissions
To obtain permission to re-use copyrighted material from Morgan & Claypool Publishers, please contact info@morganclaypool.com.

ISBN   978-1-6270-5497-3 (ebook)
ISBN   978-1-6270-5496-6 (print)
ISBN   978-1-6270-5689-2 (mobi)

DOI   10.1088/978-1-6270-5497-3

Version: 20141201

IOP Concise Physics
ISSN 2053-2571 (online)
ISSN 2054-7307 (print)

A Morgan & Claypool publication as part of IOP Concise Physics
Published by Morgan & Claypool Publishers, 40 Oak Drive, San Rafael, CA, 94903, USA

IOP Publishing, Temple Circus, Temple Way, Bristol BS1 6HG, UK

For Sara: my mate, my muse, my support, my critic, my friend.

Contents

Preface

I am *The Physicist*! Since 2006 I have run a web site, www.AskThePhysicist.com, where I answer questions about physics. The site is not intended for answering highly technical questions; rather the purpose is to answer, with as little mathematics and formalism as possible, questions from intelligent and curious laypersons. For several years before my retirement from the University of Georgia I ran a similar Q&A site for the Department of Physics and Astronomy there. Over the last decade I have answered more than 4000 questions on line and uncounted more by brief email replies. I have found this very rewarding because it is an extension of my more than 40 years experience teaching and because I learn something new almost every day. The questions I receive reveal what aspects of physics interest people and what principles they do not grasp. They reveal a wide-spread thirst to understand how physics describes, on many levels, how our Universe works. It is gratifying that the site has on the order of 50–100 000 visits per month, far more than the number of questions asked; I interpret this to mean that there are many visits by people who simply like to read and learn.

This first book is about classical mechanics. Usually 'classical' calls to mind Newtonian mechanics and that is indeed where modern physics started. Since Newtonian mechanics provides the basis for so much of physics, it is logical that it should be the subject of the first book. Today, though, classical mechanics has come to include the theory of special relativity; after all, special relativity is the correct mechanics to which Newtonian mechanics is only an excellent approximation for most aspects of our everyday lives. The first section in each chapter of the book will consist of an overview of what I consider to be the bare-bones introduction to the material. To avoid having this read like a textbook, I mainly give the overview for one-dimensional situations, avoiding vector formalism which would at worst frighten away those with limited knowledge of physics or mathematics, or at best leave them yawning. I will fill in some gaps in the overview sections in the appendices; I will indicate when a specific appendix would be helpful in understanding the answer to a specific question.

So, if you, the reader are coming to this book with little or no physics knowledge, these overviews are for you; the overviews are what you absolutely must have to get anything from the book. And, since this is not a textbook and there is no way to include all the details and subtleties of physics, there will be the occasional question and answer which you will not really understand or which will require you to do a little research on your own. If you are coming to the book with prior physics knowledge, you might want to skip the overviews, although there is always the possibility that you could gain insight or benefit from reviewing the basics.

The bulk of the book is devoted to sections which will contain mainly categorized groups of Q&As from the web site, sort of a *Best of Ask the Physicist*. Enjoy and learn!

Author biography

F Todd Baker

'The Physicist' is F Todd Baker. He received AB and MA degrees from Miami University and a PhD degree from the University of Michigan. His area of research is nuclear physics and he has published more than 70 publications in refereed journals as well as made numerous presentations at conferences and workshops. He has more than 35 years of college and university teaching experience. In 2006 he retired from the University of Georgia where he taught and performed nuclear physics research for 32 years. Previously he held a postdoctoral research associate position at Rutgers University and teaching positions at Carroll College (Wisconsin) and St Lawrence University. He now lives in Athens, Georgia with his wife Sara in a 100-year-old house mainly restored by him and decorated and landscaped by her. He has four beloved children aged 18–44 years. He enjoys bicycling around town, playing violin, cooking and baking, outdoor activities, DIY projects, film, music of many genres, working puzzles, reading mainly European murder mysteries, and hanging out in coffee houses. His *Curriculum Vitae* may be seen at http://www.ftoddbaker.com/cv.html.

Acknowledgements

With the internet at my fingertips and a few classic texts I have held on to, I can usually find answers to any reasonable layman's question in a few minutes. Over the years, several colleagues in the Department of Physics and Astronomy at the University of Georgia have helped me with subtle questions outside my expertise set. I am particularly indebted to the ROMEOs (retired old men eating out) with whom I frequently discuss interesting *Ask the Physicist* questions at our Wednesday lunches.

Chapter 1

Newtonian mechanics

1.1 Overview

Question: If a curling stone weighs 20 kg and is traveling at a speed of 0.5 m s^{-1}, with how much force did the curler throw it?

Question: What force is exerted when a 300 lb man falls 3 ft?

Question: I don't completely understand Newton's third law of motion. It says for every action, there is an equal and opposite reaction, but when we apply force to a book, why doesn't the book apply the same force to us? And why are we able to push the book wherever we want, if, according to third law of motion, the book should also have an equal reaction force?

Question: If a bullet was traveling at 823 m s^{-1} and hit an object that stopped it dead, how much force would be exerted on the target?

Every day I get questions like these, questions which say to me 'I have a feeling for what force is, a push or a pull, but I have no idea how forces are related to the motion'. In the 18th century Isaac Newton (1642–1727) conceived three simple laws which tell us how we can understand how forces affect the physical world. In order to understand how the world works, not to mention be able to read this book, we must understand these three laws.

1.1.1 Newton's first law

Imagine a book sitting on a table. It is at rest. There are two forces on it, its own weight (the force of the Earth pulling it down) and the force of the table keeping it from falling to the ground. Newton's first law simply states that, because the book is at rest, the magnitude of the weight (pointing down) must be equal to the magnitude of the table force (but pointing up), or, to put it more elegantly, the net force is zero. If questioner #3 above were pushing with a force just right so that the book moved with constant speed across the table, there would be two new forces on the book, the pusher pushing and the table resisting (called friction); but all forces (now four of them) on the book still add up to zero. An object which is at rest or moving with constant speed in a straight line is said to be in equilibrium. Newton's first law can be expressed as follows. *The net force on an object in equilibrium is zero.* This law may seem obvious today, but it was revolutionary when Newton first stated it. Before Newton it was assumed that the natural state of an object was to be at rest and, in order to keep something moving, there had to be a force pulling or pushing it.

When we get to Newton's second law, the first law will seem to be an unnecessary special case of the second. But the first law plays a much more important role than that. Suppose we ask the question 'Is a law of physics always true?' The answer, perhaps surprising, is no; there are usually conditions under which a law is true. Thinking about the first law, imagine that you are inside an accelerating car. You are at rest *inside* the automobile (in equilibrium according to the first law) but you feel the seat back pushing forward on you, an unbalanced force. Therefore Newton's first law is untrue in this automobile. Whenever you find that Newton's first law is true you are in what is called an *inertial frame of reference*. This is the more important role played by the first law, as a test whether Newton's laws are true laws for you.

When you have found one inertial frame, you have found them all because any frame which moves with constant velocity relative to another is also an inertial frame. (This is proved in the appendix G.) Any frame which accelerates relative to an inertial frame is not one. Because the Earth rotates and revolves around the Sun, it is accelerating (see appendix D) and not an inertial frame. Fortunately, the accelerations involved are small enough to be negligible for many examples in everyday life. Later in the book, though, there will be examples of physics in non-inertial frames.

Incidentally, inertial frames get their name from an alternative name of the first law, the *law of inertia*. Inertia means unwillingness to change and the first law says that you need to push or pull on something at rest or moving with constant velocity to change its motion.

1.1.2 Newton's second law

In order to discuss the second law, a brief detour to discuss units is imperative. Whole books have been written on this topic and I will be brief, assuming the reader already has a good sense of how we normally measure length, mass and time. I will usually use SI units, mass is a kilogram (kg), length is a meter (m), and time is a second (s). These are all operationally defined in rather complicated ways, but it is sufficient to simply think of a meter stick and a stopwatch for almost everything which will be discussed in this book. 1 kg = 1000 grams and 1 gram is the mass of 1 cm^3 of water; or, you might like to think of a kilogram as having a weight of about 2.2 lb if you are in a country which uses imperial units. Note that I have not talked about how force is measured; that is part of what the second law is all about.

A brief discussion of acceleration is also in order. Everyone is comfortable with what velocity is, the distance traveled divided by the time to travel it; a scientist would call this the rate of change of position. Almost nobody, in my experience, is really comfortable with what acceleration is. It is simply the rate of change of velocity. A dropped ball, for example, gains about $10\,\mathrm{m\,s}^{-1}$ in velocity for each second it falls; after 1 s it is falling with a speed of $10\,\mathrm{m\,s}^{-1}$, after 2 s with a speed of $20\,\mathrm{m\,s}^{-1}$, $30\,\mathrm{m\,s}^{-1}$ after 3 s, etc. The acceleration is $a = 10\ (\mathrm{m\,s}^{-1})\,\mathrm{s}^{-1} = 10\,\mathrm{m\,s}^{-2}$.

To discover Newton's second law you must interact with nature. We will never find physical laws by just sitting at a desk and thinking; we must make measurements which tell us how things happen. Newton's second law is about how exerting a force on a mass changes its motion, i.e. how force, mass and acceleration are related. The experiment I propose is pretty simple: push or pull on a mass with a force and measure the acceleration. First vary the mass and hold the force constant, then vary the force and hold the mass constant. Hopefully, the resulting data will lead to some general law. But there is a problem. At this stage, force is a qualitative concept, a push or a pull, and if we cannot measure it then how can we vary it or even hold it constant? But, I can imagine having a machine which always exerts the same force. I could push with my hand using the muscles in my arm with very roughly the same force each time. Or, I could attach a spring to the mass and always pull with a force such that the spring was always stretched by the same amount, an improvement over my hand/arm machine. Oh, and I will call the force my machine exerts 1 baker (B). So, let's do the first part of the experiment with one of my constant-force machines, pulling with a force of 1 B, on various masses. The data might look like the graph shown in the left panel of figure 1.1. Note that this makes sense because if the mass is large, then the acceleration is small and vice versa. The problem is that it is difficult to quantify the relationship between the two variables because the data lie on a curve, not a straight line. Suppose that, instead, we plot the acceleration as a function of the reciprocal of the mass (1/mass) as shown in the right panel of figure 1.1. This gives us a straight line which means the acceleration is proportional to the reciprocal of the mass, $a \propto 1/m$. Next, we hold the mass constant while we vary the force, first using one 1 B machine, then two, then three, etc. We would find that two gave twice the acceleration as one, three triple the acceleration, etc. In other words the acceleration is proportional to the force, $a \propto F$. Simple algebra says that therefore $a \propto F/m$. To me, this is Newton's second law, a statement of experimental facts—acceleration is proportional to force and inversely proportional to mass.

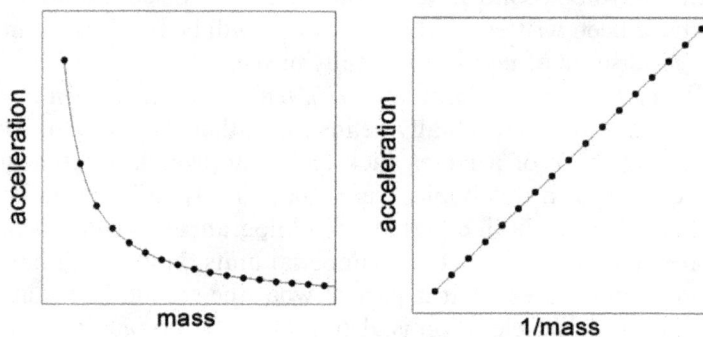

Figure 1.1. Typical data, in arbitrary units, showing the relationship between mass and acceleration for a constant force.

Since it is much more convenient to convert this to an equation, we introduce a proportionality constant C, $a = CF/m$. The choice of C determines how we will measure force. The most clever choice, of course, would be $C = 1$ resulting in $F = ma$, the way we usually see the second law written. One unit of force, called a newton (N), is that force which, when applied to an object with a mass of 1 kg, results in an acceleration of $1\ \mathrm{m\ s^{-2}}$, $1\ \mathrm{N} = 1\ \mathrm{kg \cdot m\ s^{-2}}$ and is approximately 0.225 lb. In conclusion, the second law is both a statement of an experimental fact and a definition of a unit of force.

1.1.3 Newton's third law

The third law essentially says that forces in nature always appear in pairs and that *if object A exerts a force on object B, object B exerts an equal and opposite force on object A*. I will refer to those two forces as a Newton's third-law pair and they always add up to zero. If you think about it, you will see that an alternative way of stating the third law is that *the net force on an isolated system of interacting objects is zero*, where an isolated system is one which has no forces acting on it other than the forces among its members.

There is often great confusion surrounding the third law. Carefully note from my first statement of the third law that the forces of a third-law pair are never on the same object. One of the questions cited at the beginning of this chapter wondered how we could move a book across a table since the action and reaction force always cancel out. But only one of those forces is on the book and only forces on the book determine how the book moves. For the same reason, we should not make the mistake of identifying equal and opposite forces automatically as third-law pairs. For example, the weight of a book sitting on a table points down and the force the table exerts on the book is equal but points up; these are equal and opposite because of the first law, they have nothing to do with the third law.

1.1.4 Linear momentum

Newton, in his landmark book *Philosophiæ Naturalis Principia Mathematica*, did not write the second law as $F = ma$, rather he said that the rate of change of motion is equal to the force. It will be important to understand what this means if we are to understand many of the Q&A examples in this book. Recall that the acceleration is the rate of change of velocity; it is customary in mathematics and physics to write this as $a = \Delta v/\Delta t$ where Δv is the change in velocity and Δt is the elapsed time. For example, if the velocity increases from 4 to $8\ \mathrm{m\ s^{-1}}$ over a period of 2 s, the acceleration is $(8 - 4)/2 = 2\ \mathrm{m\ s^{-2}}$. So now we can write $F = ma = m(\Delta v/\Delta t) = \Delta p/\Delta t$ where $p = mv$ is what Newton meant by the 'motion'; p is called the *linear momentum* today. (Note that since m is constant in $F = ma$, $m\Delta v = \Delta(mv)$.) If the net force on a collection of objects is zero, the rate of change of linear momentum must be zero— linear momentum never changes! This is called *conservation of linear momentum*. Conservation principles are extremely useful in physics. Notice that conservation of linear momentum implicitly invokes the third law since the net force on an isolated system must be zero.

1.1.5 Energy

The mathematics behind the idea of energy is more complex and will be handled in appendix A of this book for the interested reader. The idea is that if a force is exerted on an object as it moves through some distance, work is done on the object which changes its energy. For many situations, the force is constant and along the path of the object so that the work can be written simply as $W = Fs$ where W is the work and s is the distance traveled. Now, what changes if you do work on an object? Well, that is really an easy question to answer qualitatively if you understand the second law—if you push in the direction it is moving it speeds up and if you push opposite the direction it is moving is slows down. In other words, force causes acceleration which means either speed up or slow down in physics, so what changes is speed. Without any derivation (see appendix A), here is the way that speed changes: $W = \Delta(\frac{1}{2}mv^2) = \frac{1}{2}mv_{\text{final}}^2 - \frac{1}{2}mv_{\text{initial}}^2$. This is often called the work–energy theorem. When you do work, you change the quantity $K = \frac{1}{2}mv^2$ which is called the kinetic energy of the object. Note that if there is no work done on a system its kinetic energy never changes. Again, we have discovered a conservation principle, conservation of energy, which states that *a system on which no forces do work will have its total energy constant.* Something called potential energy is useful, but for the most part it will not be needed for this book. I will briefly discuss and define potential energy in appendix A and write the potential energy for weight, mgy.

Finally, the unit to measure energy and work is the joule (J), $1\ \text{J} = 1\ \text{kg·m}^2\ \text{s}^{-2}$. You are also probably familiar with the unit of power, the rate at which energy is used or created, the watt (W). $1\ \text{W} = 1\ \text{J s}^{-1}$. A 100 watt light bulb consumes 100 J of energy each second.

1.1.6 This is all wrong!

As we shall learn in the second part of this book, Newtonian mechanics, as framed above, is only an excellent approximation to the true classical mechanics, special relativity. $F = ma$ is wrong, $K = \frac{1}{2}mv^2$ is wrong, $p = mv$ is wrong. But, this need not bother us here in chapter 1 because only when speeds become comparable to the speed of light, $c = 671\,000\,000$ mph, might you notice that Newtonian mechanics is not right. Interestingly, Newton's expression for the second law, $F = \Delta p/\Delta t$, is correct after linear momentum is slightly redefined.

1.2 Newton's laws misunderstood

At the beginning of section 1.1 were listed several questions which demonstrate how Newton's laws are often misconstrued. Let us now look at some of those questions along with the answers.

> **Question:** Can you explain to me what exactly keeps molecules moving? With no energy being added, they should just eventually stop, shouldn't they? Where does this energy that keeps them moving come from? In the end, does it all come down to radiation from the Sun?

Answer: You have fallen into one of the most common traps regarding misunderstanding how the Universe works. Newton's first law states that an object which experiences no net force will continue to move with constant speed in a straight line. What this means is that if something is moving and nothing is pushing or pulling on it, then you do not have to do anything to keep it moving. In terms of energy, if something has a certain amount of energy, then it will retain that energy until some external agent changes it; this is called conservation of energy. I am not sure what you have in mind with your question, but probably the molecules moving around in a gas. As you probably know, the temperature of a gas is a measure of the average kinetic energy per molecule. If the gas is in thermal equilibrium with the walls, then when a molecule hits the wall it rebounds (on the average) with the same kinetic energy it had beforehand. You don't have to do anything to keep it moving. Incidentally, if Newton's first law were not true we would never have sent probes to the distant planets like Saturn and Jupiter or even the close ones like Mars and Venus. The reason is that if we had to keep the probe moving by burning an engine the whole way we could never carry enough fuel. What actually happens is that we burn up almost all the fuel escaping the Earth and acquiring a high speed and then we just turn off the engines and coast the rest of the way.

Question: If a curling stone weighs 20 kg and is traveling at a speed of $0.5\,\mathrm{m\,s^{-1}}$, with how much force did the curler throw it in N?
Answer: You cannot determine the force needed to give a particular mass a particular speed. Just to make that plausible, suppose you push on the 20 kg stone with a force of 2 N for 1 s; surely it will have a different result than if you push on the 20 kg stone with a force of 2 N for 2 s. There are two (in the end, equivalent) ways you can think about this problem:
- The impulse delivered by a force F in a time t is Ft. The linear momentum of an object with mass m and speed v is mv. The change in momentum is equal to the impulse and so, if the object starts at rest, $Ft = mv$. For example, in your case $Ft = 10\,\mathrm{kg\cdot m\,s^{-1}}$ so you could push with a force of 10 N for 1 s.
- The work done by a force F pushing over a distance s is Fs. The kinetic energy of an object with mass m and speed v is $\frac{1}{2}mv^2$. The change in kinetic energy is equal to the work and so, if the object starts at rest, $Fs = \frac{1}{2}mv^2$. For example, in your case $Fs = 2.5\,\mathrm{kg\cdot m^2\,s^{-2}}$ so you could push with a force of 10 N for a distance of 0.25 m.

In both cases, be sure to note that what the force is depends on how long or far it is applied.

Here the questioner thinks that the speed something acquires depends on how hard you push it, so that if you know its speed you know how hard it was pushed. But, it is acceleration, not speed, to which force is related and, as the answer shows, a small force over a large distance or time has the same effect as a large force over a small distance or time.
Sometimes the question can be answered if particular constraints are placed on it.

Question: We are working to produce a safety harness and the strap material we are using has a maximum Newton rating—we were trying to get an idea of what Newton rating would be needed to support a 300 lb man if he fell 3 ft. Being hunters (tree stand safety harness)—perhaps we are wording the question incorrectly.

Answer: What matters is how long it takes the falling guy to stop. The mass of a 300 lb guy is about 130 kg, the acceleration of gravity is about 10 m s^{-2}, and so the weight of the guy is about 1300 N. You need that strong a strap just to hang him there at rest. If he falls 3 ft (about 1 m) he will be going about 4.5 m s^{-1}. So, let's call F the average force needed to stop him and t the time it takes him to stop; I reckon that $F \approx 130(10 + (4.5/t))$. For example, if he takes $\frac{1}{4}$ s to stop, $F \approx 3600 \text{ N} \approx 809 \text{ lb}$ to stop him. The straps are probably pretty unstretchy, so your best bet would be to make the harness out of a stretchy material because, don't forget, the bigger F is the more it is going to hurt during the stop.

Note that the average acceleration as he is stopping is $\Delta v/\Delta t$ which is where the $4.5/t$ came from; $a = 4.5/t$. So the strap needs to do two things, hold up the weight (mg) and provide the acceleration (ma) where $m = 130$ kg.

Question: If a bullet was traveling at 823 m s^{-1} and hit an object that stopped it dead how much force would be exerted on the target?

Answer: Here is the question which I get in one form or another which indicates how poorly understood the concept of force is! You cannot get the force because it depends on how quickly the bullet stops. If you mean by 'stopped it dead' that it stops instantaneously, then the force would be infinite. The average force is the change in momentum (mass times velocity) divided by the time to stop. So, you need also the mass of the bullet. Suppose the bullet had a mass 0.02 kg, then the change in momentum (0.02×823) is about 16 kg·m s^{-1}. If it stops in 0.01 s the average force is $1600 \text{ N} = 360 \text{ lb}$, if it stops in 0.001 s the average force is $16\,000 \text{ N} = 3600 \text{ lb}$.

This next question comes up often. Newton's third law says all forces have an equal and opposite mate. Why don't they all add up to zero so that nothing ever moves?

Question: If action and reaction are always equal in magnitude and opposite in direction, why don't they always cancel one another and leave no net force to accelerating a body?

Answer: Newton's third law states that if one object exerts a force on a second, the second exerts an equal and opposite force on the first. Therefore, the 'action/reaction' forces are never exerted on one body. If you select a body to study, its motion is determined only by the forces exerted on it, not by forces exerted by it. Students often make mistakes with this 'action/reaction' thing because they tend to identify any pair of equal and opposite forces as being an 'action/reaction' pair. For example, a 1 lb book sitting on a horizontal table has two forces on it, its 1 lb weight pointing down and a force of 1 lb which the table exerts up on it (usually called the normal force); these have nothing to do with Newton's third law but are equal and opposite because the book is in equilibrium and the force the table exerts is therefore required to be 1 lb up. If we now look at the table, the book exerts a 1 lb force down on it because of Newton's third law; the 'action/reaction' pair is the force the table exerts on the book and the force which the book exerts on the table. Lots of novice physics students want to say that the weight of the book is the 1 lb force down on the table—this is totally false since this is a force on the book, not the table.

1.3 Air drag

If you have ever taken an introductory physics course, an often encountered phrase is 'neglecting air drag' or 'neglecting air friction'. It is often a good idea to make approximations when you are just starting to learn something, deal with the relatively simple cases first. If you have a marble that you drop from 3 m, it is a very good approximation to neglect air drag, but what if you drop a cotton ball from 3 m or a marble from 1000 m? There are many examples of motion of objects where air drag is important and many questions I get are about cases where air drag is important.

Most problems of interest for objects moving through air can be well approximated as encountering a drag force F_d proportional to the square of the speed v of object, $F_d \propto v^2$. In fact, there is a fairly good expression for the proportionality constant necessary to make this an equation: $F_d = Cv^2 = (C_d A\rho/2)v^2$ where C_d is the drag coefficient which depends on the shape of the object, A is its cross-sectional area, and ρ is the density of the air. A reasonable approximation if SI units are used is for $C = \frac{1}{4}A$. An important thing to recognize if air drag is important is that there is what is known as a 'terminal velocity', v_t, the speed which an object moving through the air tends toward. If dropped, it speeds up to v_t, if projected at a higher speed it slows to v_t. It is easy to calculate v_t because it is the speed where the drag force (up) becomes equal to the weight force (down), $Cv_t^2 = mg$ or $v_t = \sqrt{(mg/C)}$. So, contrary to the simple Galileo-story result that all objects fall with the same acceleration, if two objects having identical shapes and sizes are dropped, the more massive one wins because the terminal velocity is larger because it is proportional to the square root of the mass.

One thing to keep in mind as we look at a few examples is that whenever you include air drag, your calculation is approximate. The details of air drag are very complicated and best done numerically with big computers if you are designing an airplane!

Often, questions involving air drag are about sports as these next three questions are.

Question: How much does a lacrosse ball (2 inch diameter) slow down (horizontal velocity only) if thrown at 80 mph from the instant it is released until it reaches a point 10 m away, taking into account air resistance.

Answer: I prefer to work in metric units so 80 mph is about $v_0 = 35$ m s^{-1} and the diameter is about $D = 6$ cm $= 0.06$ m. I will also need the mass of a lacrosse ball which I looked up to be about $m = 0.15$ kg. Now, for a ball of this size traveling through air with this velocity, the air resistance force is proportional to the square of the velocity. Therefore Newton's second law is of the form $-Cv^2 = ma = m(dv/dt)$ where C is a constant which can be calculated approximately as $C = 0.22D^2$ for a sphere in air. Therefore we must solve the differential equation $(dv/dt) + 0.00079v^2 = 0$. (I completely ignore gravity because the ball starts with zero velocity in the vertical direction and flies for only a very short time.) If you know differential equations, then this is not particularly difficult to solve. I will do that later. For starters, however, it is instructive to make a reasonable approximation and see what we get. I am going to say that I expect, over so short a distance as 10 m and starting with such a large initial velocity, that the acceleration will not change much. So I will say that the acceleration at the beginning, $a_0 = -0.00079 \times 35^2 = -0.97 \text{ m s}^{-2}$, does not change much over the flight. So we have a uniform acceleration problem and we can say $x = v_0 t + \frac{1}{2}a_0 t^2 = 10$ and solve for t; I find that $t = 0.29$ s. Finally, we can obtain the estimated final velocity, $v = v_0 + a_0 t = 35 - 0.97 \times 0.29 = 34.7 \text{ m s}^{-1}$. So the ball loses about 0.9% of its initial velocity.

For anyone interested in the exact solution of the differential equation, here it is. The solutions to the equation are $v = v_0/(1 + kt)$ And, $x = (v_0/k)\ln(1 + kt)$ where $k = Cv_0/m$. Solving these I find that $t = 0.29$ s and $v = 33.2 \text{ m s}^{-1}$. So, only about 5% of the velocity is lost.

The previous question was done in two ways. One very important thing to know in science is how to make approximations to make a problem more manageable without getting incorrect results. One of the things which makes air drag problems tricky is that the force depends on the speed and so the acceleration does also. In this problem I suggested that, since the time it takes a fast lacrosse ball to go a short distance must be really small, the velocity, and therefore the acceleration, does not change very much. Then we can use the equations for uniform acceleration to solve the problem, much more familiar to many of you than the more difficult solution to the differential equation. And the exact and approximate solutions give you pretty much the same result.

This next question is about baseball. It is well known that a curve ball happens because of air drag but I had not realized how much a ball slows down in the brief time it takes a fastball to reach the plate. Here I use the 'exact' solution (in quotes because all air drag calculations are approximate).

Question: Based on physics, is a 90 mph fastball slower or faster than a 95 mph fastball? At work we are trying to determine if the 95 mph fastball loses energy faster than a 90 mph fastball. Your answer is greatly appreciated.
Answer: You are asking two questions; if a 95 mph ball loses energy faster than a 90 mph fastball (it does) and if the one which starts out faster ends up slower (it does not). For the details of the following, see the earlier lacrosse ball answer. Following the (exact) solution in that earlier answer, I find that the 95 mph ball reaches the plate in 0.47 s and arrives at the plate with a speed of about 80.8 mph. The 90 mph ball reaches the plate in 0.50 s and arrives at the plate with a speed of

about 76.3 mph. So, each loses about 14 mph with the faster ball losing a bit more. This surprised me but I found another reference saying that something like 10 mph is what is lost, so my calculations are reasonable. So they do not lose energy significantly differently (the faster pitch lost more speed in a shorter time so its average rate of change of speed was indeed bigger). (I used 3 inches for the diameter, 0.145 kg for the mass, and 60 ft 6 inches for the distance to the plate.) There is certainly no way that one could characterize a 95 mph fastball as slower than a 90 mph fastball.

Often I am called on to settle arguments. Here is an example involving air drag and the sports of tennis and badminton.

Question: A friend of mine and I have an argument over which is the faster sport, tennis or badminton. The criterion is how long it would take to serve a tennis ball/shuttlecock from one side of an Olympic sized tennis/badminton court to the player waiting on the other side assuming that both are standing on the out of bounds line. We are assuming ideal conditions and that the players in both cases are equally strong and fast.

Answer: You may not realize it, but your question is mostly about air drag on projectiles. I seem to get more questions about air drag than just about anything else except maybe variations of the twin paradox. Maybe that is because it is perhaps the most important phenomenon mostly swept under the rug in most elementary physics courses. There are several instances of earlier questions involving baseballs and lacrosse balls which are very similar to this one. For high speed projectiles, air drag is very important; e.g. a 100 mph baseball loses about 10 mph by the time it crosses the plate. Approximations have to be made to quantify the situation you are interested in, but I feel the results I will present are pretty close to what happens on the court. The approximations are:

- I neglect gravity because the times involved are sufficiently short that the ball/shuttlecock will not fall far or very much change its vertical speed.
- I assume that the drag is proportional to the square of the speed—twice the speed, four times the force of drag. This is an excellent approximation for these speeds, these objects.
- The form of the force I use is $F \approx \frac{1}{4}Av^2$ where A is the cross-sectional area presented to the wind. Here $A = \pi R^2$ where R is the radius of the ball or the outer circle of the feathers. This probably slightly overestimates the force for the tennis ball (whose 'hairs' have the function of decreasing the drag) and underestimates it for the shuttlecock (whose 'feathers' are designed to increase drag).
- Data for tennis:
  $v_0 = 73 \text{ m s}^{-1} = 163 \text{ mph}$
  $R = 0.032 \text{ m} = 1.26 \text{ in}$
  $m = 0.057 \text{ kg} = 2 \text{ oz}$
  back line to back line distance: 24 m
- Data for badminton:

$v_0 = 92 \, \text{m s}^{-1} = 206 \, \text{mph}$

$R = 0.025 \, \text{m} = 1 \, \text{in}$

$m = 0.005 \, \text{kg} = 0.18 \, \text{oz}$

back line to back line distance: 13.4 m

I used the fastest recorded serves for the velocity off the racquets, v_0. If you integrate $F = ma$, you get the following solutions: $v = v_0/(1 + kt)$ and $x = (v_0/k) \ln(1 + kt)$ where $k = \frac{1}{4}Av_0/m$. Here are the results:

- The tennis ball takes 0.39 s to travel the distance, arrives with a speed of $52 \, \text{m s}^{-1}$ (116 mph), a loss of $21 \, \text{m s}^{-1}$ (47 mph) or 29%.
- The badminton shuttlecock takes 0.30 s to travel the distance, arrives with a speed of $25 \, \text{m s}^{-1}$ (56 mph), a loss $67 \, \text{m s}^{-1}$ (150 mph) or 73%.

I will leave it to you to argue about what these numbers tell you about which 'is the fastest sport'. According to your criterion, the shuttlecock arrives earlier but with a much lower speed. The shuttlecock starts off the fastest because it has a smaller mass and can therefore have a larger acceleration from the force from the racquet. But it slows down very rapidly mainly because of its small mass. Figure 1.2 shows the speeds as functions of time over the flight time of each.

Finally, in our exploration of air drag, here are a few questions in which terminal velocity is the focus.

Question: If I were to drop an empty wine bottle out of an airplane flying at say 35 000 ft above the ocean at 300 mph, would the bottle hit the surface of the water hard enough to break the bottle? I read somewhere something about terminal

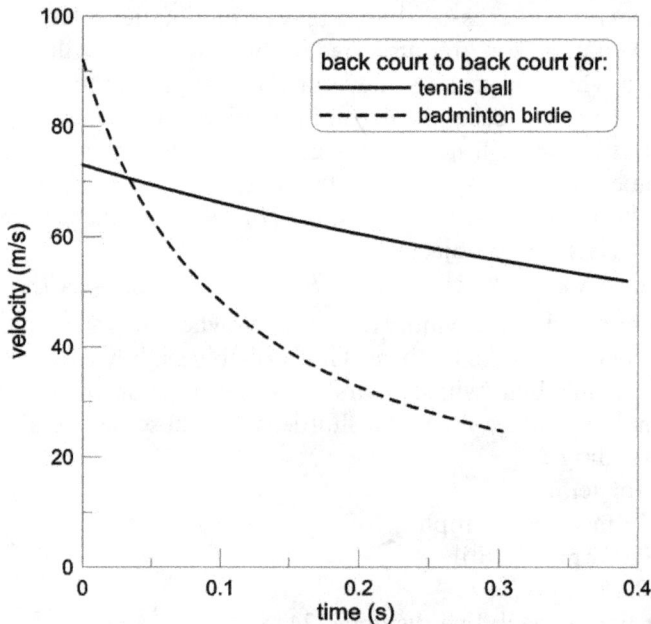

Figure 1.2.

velocity being 120 mph, so would the resistance of the atmosphere slow the wine bottle to 120 mph by the time it made impact with the ocean? And would 120 mph be enough to shatter the wine bottle, or would it depend on how choppy the seas were versus a flat water surface?

Answer: When I answer questions involving air drag and terminal velocity, I usually use the approximation that (in SI units) the force F of air drag is $F \approx \frac{1}{4}Av^2$ where A is the area presented to the wind and v is the speed. So, as something falls, the faster it goes the greater the drag force on it so that, eventually, when the drag equals the weight, the object will be in equilibrium and fall with constant speed. Since the weight W is mg where m is the mass and $g = 9.8$ m s^{-2}, the terminal velocity v_t can be calculated: $\frac{1}{4}Av_t^2 \approx mg$ or $v_t \approx 2\sqrt{(mg/A)}$. So the terminal velocity depends on the mass and size of the falling object and your 120 mph is most likely not correct. Also, how it falls determines the terminal velocity since it has a much bigger area falling broadside than with the top or bottom pointing down. I figure that if it falls broadside there will be a bigger force on the fat side than the neck which will cause a net torque which will make it want to turn with its neck pointing down; so I will assume that is how it falls. I happened to have an empty wine bottle in my recycle bin which has a mass of about 0.5 kg and a diameter of about 8 cm. When I calculate the terminal velocity I get $v_t \approx 63$ m s^{-1} = 140 mph. The 120 mph number you heard was probably a typical terminal velocity of a human, and it is just coincidence that the wine bottle has a terminal velocity close to that.

It is hard to say whether it would break or not. I think probably not. Suppose that it took 1 s to stop. Then the average force on the bottle would be $F = ma = (0.5$ kg $\times 63$ m s$^{-1})/(1$ s$) = 31.5$ N ≈ 7 lb which the bottle should be able to withstand easily. I know that they say that at high speeds hitting the water is like hitting a brick wall, but if the stopping time were 0.1 s the force would still only be about 70 lb.

Question: If you shoot a bullet straight up into the air, its velocity at the very top of the trajectory is zero, even if only for an instant, as its upward velocity slows to nothing before becoming downward velocity. Downward vertical velocity then increases in the earthward direction. Would the velocity ever become dangerous if it landed on a living person? Is the weight of the bullet important? Does the atmosphere restrict the downward velocity?

Answer: A falling bullet experiences a downward force of its own weight and an upward force of air drag. The result of the air drag, which increases with speed, is to have the falling object eventually reach a maximum velocity called the terminal velocity which is determined by its weight and its geometry (which is why you can jump out of an airplane with a parachute). A .30 caliber bullet weighing about 10 grams has a terminal velocity of about 90 m s^{-1} (about 200 mph) and a .50 caliber bullet weighing about 42 grams has a terminal velocity of about 150 m s^{-1} (about 335 mph). A bullet traveling 60 m s^{-1} (about 130 mph) can penetrate the skull so, yes, a falling bullet is dangerous. Dozens of people are killed every year by celebratory gunfire.

Question: Why is it difficult to calculate the terminal velocity for a cat falling from a high roof top?

Answer: I do not know what you mean 'difficult to calculate'. We can estimate it pretty easily, but certainly not do it precisely. First of all, any calculation having to do with air friction is going to have approximations and assumptions. For something like a cat, roughly 2 kg (4.4 lb), falling, it is a very good approximation to say that the drag force is proportional to the square of the velocity. It turns out that a fairly good approximation for the force is $F = \frac{1}{4}Av^2$ where A is the area the falling object presents to the onrushing wind and v is the velocity (this is only for SI units). Since it depends on A, it depends on how the cat orients itself: if in a ball he will fall much faster than if all spread out. Suppose we take the area of a falling cat to be about $20\,\text{cm} \times 40\,\text{cm} = 0.08\,\text{m}^2$. Then the force will be about $0.02v^2$.

Now, the cat's weight is about $mg = 2 \times 9.8 \approx 20$ N. When the force of air friction is equal to the weight force down, the cat will fall with a constant velocity called the terminal velocity: $0.02v_t^2 = 20$, so $v_t = \sqrt{(20/0.02)} \approx 30$ m s^{-1} = 67 mph. If you google 'terminal velocity of a cat' you will find the number 60 mph, so my approximations were evidently reasonable. There, now, that wasn't so difficult, was it?

This last question is interesting in that, since cats have a relatively low terminal velocity, they usually survive falls from high buildings. In fact, they are more likely to survive falls from higher than seven stories because, at the time they reach terminal velocity, they instinctively relax and spread out. The next question concerns an animal with a much smaller terminal velocity.

Question: Why is it that if you blow a spider suspended by her web she floats out but then when this pendulum swings back it stops when the web is vertical and doesn't swing back and forth? Is it due to the air friction as it comes back to equilibrium or perhaps the dynamic structure of the web strand that absorbs energy that would have made the web swing back and forth?

Answer: It is caused by air drag. This is called a damped oscillator. If there were no air, the spider would swing back and forth with constant amplitude, just like a clock pendulum (apart from the little friction from bending the thread she hangs from). A spider has so little mass that her terminal velocity is very small—drop her off the roof and she will not get hurt because she quickly comes to some constant velocity because the air drag, which can be approximated as being proportional to her speed, quickly becomes equal to her weight. If the air drag is not too big, the pendulum will swing back and forth with ever decreasing amplitude; this is called underdamped. For larger drag, as in the case of your spider, she never crosses over the equilibrium and just slowly approaches the bottom of her swing; this is called overdamping. There is a third possibility called critically damped, but it is qualitatively just like overdamping, so let's not go there.

1.4 Gravity and Kepler's laws

Gravity is the force which you are most aware of. This is strange because, gravity is the weakest force in nature. How can that be? The only reason that it is so pervasive in your life is that there happen to be very huge accumulations of mass scattered throughout the Universe (like the Earth, the Sun, etc) and mass is the source of gravitational force. Newton discovered that two objects of mass M_1 and M_2, separated by a distance r, exert a force F on each other given by $F = GM_1M_2/r^2$ where $G = 6.67 \times 10^{-11}$ N·m^2 kg^{-2} is the universal constant of gravitation. Now you can appreciate what a weak force gravity is: two 1 kg masses separated by 1 m exert a force on each other of 6.67×10^{-11} N; this is about 100 times smaller than the weight of one speck of dust.

How did Newton figure this out? Astronomy was a science which was around long before physics. Amazingly accurate measurements of the positions of planets were made by the Danish astronomer Tycho Brahe and his assistant German astronomer Johannes Kepler in the 16th century, nearly a century before Newton's birth. Using these data, Kepler was able to describe how the planets move in their orbits using his now-famous three laws; the laws, however, were purely empirical, meaning that they resulted from just describing data, not any physical principle. Newton's triumph was that his law of gravitation was able, along with his three laws of mechanics, to explain Kepler's laws.

Finally, note that near the Earth's surface the force W on a mass m is approximately $W = m(GM/R^2) \equiv mg$ where $M = 6 \times 10^{24}$ kg and $R = 6.4 \times 10^6$ m are the mass and radius, respectively, of the Earth. So g, the acceleration due to gravity, is $g = 9.8$ m s^{-2}.

A reasonable question was how Newton could know G. He never did.

Question: How was the value of $G = 6.67 \times 10^{-11}$ derived? How did Newton get the value of the constant G?

Answer: G is a fundamental constant of nature, it cannot be derived. Newton did not get the value of G, the best he could do was get the product GM where M is the mass of the Sun. He shows that $GM = 4\pi^2 a^3/T^2$ where a is the semimajor axis of the orbit of a planet and T is the period. This is a derivation of Kepler's third law and is the real triumph of Newton. It was 70 years after Newton's death that the first measurement of G was made by Cavendish. You can also get $M_{Earth}G = R^2 g$ where M_{Earth} is the mass of the Earth, R is the radius of the Earth, and $g = 9.8$ m s^{-2}; so Newton could have found the ratio M_{Earth}/M without knowing G.

You have probably heard the legendary tale of Galileo dropping balls from the Tower of Pisa, finding them all falling in the same time. It is a question frequently submitted to *Ask The Physicist*.

Question: Why do two objects of different masses reach the ground at the same time and what are the factors that affect their motion?

Answer: The motion of a mass m is determined by Newton's second law, $F = ma$, where F is the net force on m and a is its acceleration. A mass in free fall (no air friction) has only one force on it, its own weight W which is the force with which the Earth pulls on it. It turns out that the weight is proportional to the mass, that is $W = mg$ where g is a proportionality constant called the acceleration due to gravity. So, if you have two masses, m and M, you can calculate their accelerations, a and A, respectively: $A = W/M = g$ and $a = W/m = g$. Since both have the same acceleration, they fall identically. (You can see why g is called the acceleration due to gravity.) The factors which must be satisfied for this to be true are that air friction is negligibly small and the masses must be small compared to the mass of the Earth.

It turns out that there is a profound physical truth here. There are really two kinds of mass we have discovered now. One is inertial mass, the property which resists acceleration when you push on it; the other is gravitational mass, the property which allows objects which have it to create and feel gravity. The fact that different masses have the same acceleration implies that the two masses are identical. Experiments bear this out to remarkable accuracy. This is important in general relativity, the modern theory of gravity, and will be revisited in chapter 3.

I have had many variations of the next question. Above, the important role in physics history played by Kepler's laws was emphasized. Here is a question the answer to which states these laws and uses two of them.

Question: The Earth orbits around the Sun. If we stopped the Earth in orbit and then let it fall straight towards the Sun, how long would it take to reach the Sun?
Answer: The questioner sent me a bunch of data about the masses of the Sun and Earth, the radius of the Earth's orbit, and Newton's universal constant of gravitation. But, you do not need any of that stuff—all you need to know is Kepler's first and third laws and the fact that the period of Earth's (approximately circular) orbit is one year. Kepler's laws are as follows.
  - The first law states that the orbit of a planet is an ellipse with a semimajor axis a and with the Sun at one focus of the ellipse.
  - The second law states (not needed for this question) that a planet in its orbit sweeps out equal areas in equal times, so it moves faster as it gets closer to the Sun.
  - The third law states that the square of the period T of an orbit is proportional to the cube of its semimajor axis, $T^2 \propto a^3$.

The Earth's orbit is very nearly circular and a circular orbit has a semimajor axis equal to the radius of the circle, so $a_1 = R_O$ where R_O is the radius of the Earth's orbit; the eccentricity of a circle is 0. The other extreme is an ellipse with eccentricity 1 which is a straight line from the Sun to the Earth and so the semimajor axis for a 'dropped Earth' is $a_2 = R_O/2$. (To help you visualize this, figure 1.3 shows an elliptical orbit very close to the straight line orbit; just squeeze it a little bit flatter.) If we can cleverly deduce the period of this orbit, one-half that period will be the answer to your question. Using the third law,

$$T_1^2/T_2^2 = a_1^3/a_2^3 = R_O^3/(R_O/2)^3 = 8$$

$$T_2 = T_1/\sqrt{8} = 0.354 \text{ years.}$$

So, the time to go half a period is 0.177 years = 64.6 days.

Figure 1.3.

To actually do this problem by brute force, integrating Newton's second law for the gravitational force which changes as the Earth falls into the Sun, is very tricky and certainly beyond the understanding of the average interested layperson. This problem emphasizes, once again, an important thing to learn in science—it is often most illuminating to make reasonable approximations to difficult problems. The solution above is not actually perfectly accurate, it treats the Sun and the Earth as point masses which is certainly not true. Also, the Earth would have infinite velocity when it turned around in its straight line orbit and its acceleration would be infinite at that time, obvious nonsense. But, if we realize that the real end of the trip is at the surface of the Sun and that distance is tiny compared to the total distance to the Earth's orbit, most of the time must have been spent in the fall to the surface, the tiny amount farther being a negligible contribution. I show in appendix B that the time neglected is less than about 9 min of the 64.6 day total. As we shall see in chapter 2, for Newtonian mechanics to be applicable the speed must be much smaller than the speed of light and in this case the Earth has about 0.4% the speed of light when it gets to the surface, again shown in appendix B.

The following question is also a frequent one. Variations on 'what happens when I drop a stone in a hole drilled through the Earth?'

Question: A sky-diver is falling toward Earth. A tunnel has been previously excavated completely through the Earth at exactly the location of the skydiver's landing. He continues his dive through the tunnel without touching the sides of the tunnel. I believe that Newton would have had him stop at the Earth's core. Where would Einstein have him stop?

Answer: First, since this is clearly an idealized problem, let us neglect air friction (which is, of course, not negligible because the sky-diver has a terminal speed before he hits the ground). Until he enters the tunnel he is accelerating with a constant acceleration down. When he enters the tunnel, he experiences less and less force as he goes deeper because there is less and less of the Earth pulling on him (all of the Earth outside him exerts no force) until finally at the center he has zero force on him but he has his highest velocity of the whole trip since he has been speeding up the whole time. Now as he moves away from the center he slows down. When he re-emerges at the other end of the tunnel he has exactly the same speed as he had when he entered it. He continues until he reaches the altitude from which he originally jumped at which point he turns around and begins the process all over again. Newton and Einstein would both agree on this. If air friction were included, he would not go as far and if he happened to stop at the center of the Earth, he would stay there forever. If the air friction were included, the general solution to the problem would be that he would oscillate back and forth going less far each time until he finally stopped in the center. There is an interesting aspect of this problem: if the Earth had its mass uniformly distributed through its volume (it does not), when the sky-diver is inside the tunnel, he moves exactly like he were a mass on an ideal spring.

The next question has to do with golfing on the Moon.

Question: My son has a question that I can't seem to find an answer to researching on the web. What force is behind a golf ball when hit on the Moon? I appreciate your time on answering this for my son.

Answer: I think there is some confusion about what force is here. Here are all the forces a golf ball experiences here on Earth:
- The club, traveling with some speed, hits the ball and exerts a contact force on it for a very short time but it is a very big force and it results in the ball

acquiring a very large velocity. As soon as the ball leaves the club, there is no force 'keeping it moving'. If there were no other forces, it would keep going forever with the speed with which it left the club.

- Once it is started, gravity pulls down on it which is the force which eventually does bring the ball back to the ground.
- As it flies through the air, it experiences air drag which can be very complicated. Essentially, it is a force trying to slow it down and the bigger the speed is the bigger this force is.
- If it happens to be spinning, the air drag can act asymmetrically so that the ball curves. This is what is called a hook or a slice in golf (depending whether it curves left or right, respectively, for a right-handed golfer).
- Of course, when it hits the ground, it experiences forces from the ground which ultimately bring it to rest.

What is different on the Moon?
- If the club is the same club with the same speed, there is no difference for this force. Therefore, the ball launches just the same as on Earth.
- The Moon is much smaller than the Earth and the result is that the gravity on the Moon is much weaker. Therefore, this force (trying to pull the ball back down) is much smaller and the ball will go a lot farther.
- Since there is no air on the Moon, there is no drag and this also results in the ball going much farther.
- The ball will not curve on the Moon, regardless of how much spin it has.
- When it hits the ground, things are about the same as on Earth except that all the forces are smaller, again because of gravity being smaller, so the ball rolls farther before it stops (also because it is going much faster when it hits the ground than it would have been on Earth).

1.5 Physics of everyday life

Physics applies to many every-day situations. Even analogies between societal issues and physics can be found, as the next question illustrates.

Question: It's a question the answer to which I wish to use as an analogy when I make talks to citizen groups regarding homelessness; and specifically in response to the complaint by some in the audience that the homeless need to just pick themselves up by their own bootstraps and stop being a burden on society. I keep trying to explain to them that once one has fallen all the way down (as opposed to just tipping over a little, or even falling to one's knees; and especially once they've slipped so through certain kinds of society's cracks), it actually takes more effort to get back up again than it took to knock the person down. (And, trust me, it does.) [*The Physicist*: The questioner wishes to compare the energy necessary to tip over a cylinder of radius R whose center of gravity is a distance h above the floor to the energy required to lift it back up.] What is the amount of energy needed to tip it over from vertical to horizontal compared to the amount of energy needed to tip it back up and make it vertical again? I'm looking for a ratio.
Answer: Figure 1.4 illustrates the situation described below. (See appendix C.) To tip it over, you have to move the center of gravity (COG) so it is above the point on the floor where the cylinder touches the floor; to do this you must raise the COG a distance $d = h[\sqrt{(1 + (R/h)^2)} - 1]$. The work necessary to do this is $W_{\text{fall}} = mgh[\sqrt{(1 + (R/h)^2)} - 1]$. If R is much smaller than h, this may be approximated as $W_{\text{fall}} \approx \frac{1}{2}mgh(R/h)^2$. The work necessary to lift it back up is $W_{\text{lift}} = mg(h - R)$. Again, if R is much smaller than h, $W_{\text{lift}} \approx mgh$. So, the ratio is $W_{\text{lift}}/W_{\text{fall}} \approx 2(h/R)^2$. For example, if $h = 5R$, $W_{\text{lift}}/W_{\text{fall}} \approx 50$; it takes 50 times the work to lift as to push over!

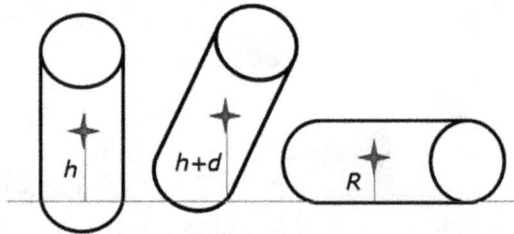

Figure 1.4. A cylinder upright, about to tip over, and fallen.

Question: Could you explain why the driver of a car must keep her foot on the accelerator to maintain a constant speed and therefore why energy is needed to maintain the car's speed?
Answer: Wouldn't it be great if we could have a car which had no energy loss? Unfortunately, the world has forces which we cannot avoid which take energy away from something moving along. These fall into the category of frictional forces: a spinning wheel has friction in its bearings which will eventually cause it to stop; an object moving thought the air has air resistance which will eventually

stop it as it moves along; the tires are not perfectly elastic and as they roll they are being continually deformed and undeformed and energy is lost. Without all these forces, we could accelerate up to speed and disengage the engine from the wheels and turn it off and just cruise. However, one can work hard to minimize these forces in the design of cars; making the cars aerodynamic, reducing the weight, and other tricks can minimize the energy we lose.

Bicycle stability involves some pretty heavy physics, but some aspects of it can be understood fairly easily as in the following question. Look for this question to come up again in section 1.6.

Question: Why, when a cyclist is turning round a bend, why does he lean inward? **Answer:** Figure 1.5 is a simplified picture of the bike and cyclist, the circle representing the center of mass. Now, the sum of all the vertically directed forces must add to zero, $-mg + N = 0$ which tells you that $N = mg$. And, the sum of all the horizontally directed forces must equal mass times acceleration, $F_f = mv^2/r$. So, given m, v and r, you now know all the forces. But you still need to know the angle of lean for the cyclist to not topple over. This is achieved by summing all torques (about the center of mass) and setting it equal to zero (so that it does not start to rotate in the plane of the page), $\Sigma\tau = 0 = NL \sin\theta - F_f L \cos\theta = mgL \sin\theta - (mv^2 L/r)\cos\theta$. And so, solving for θ, $\theta = \tan^{-1}(v^2/rg)$. (See appendices C and D.)

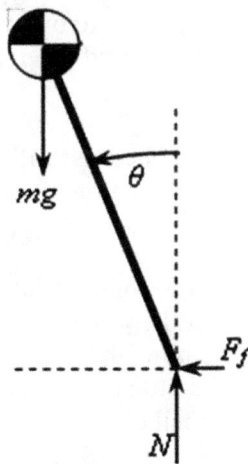

Figure 1.5. A bicycle turning a curve.

The physics of towing something is illustrated in the next two questions.

Question: If I am towing a vehicle from a standing start is there an equation for calculating the amount of force I would be using for example if I tow a vehicle that weighs twelve tonnes because it has wheels and is therefore not a 'dead weight' how do I work out how much force I would be exerting on the tow rope/ towing vehicle and also how would I factor in different gradients as it would obviously require greater force on an upslope. This came up in my workplace where our towropes are rated to three tonnes and I was trying to explain that it does not mean you could not tow a vehicle over that weight.

Answer: This is a good question to illuminate elementary Newtonian physics. Your referring to 'dead weight' really has no meaning in physics, but you apparently mean that the object can move with little friction. So, let's assume there is no friction; this is, of course, never true, but it puts an upper limit to anything I do. On level ground, any force will move the vehicle if there is no friction. What matters is how quickly you start it moving, in other words what the acceleration is. For example, suppose you have a 4 lb fish hanging on a 5 lb test line; if you pull it up slowly you will land it, but if you try to jerk it up really fast the line will break. The physical principle in play here is Newton's second law, $F = ma$ where F is the force, m is the mass, and a is the acceleration of m due to F. So, in your case, $m = 12\,\text{t} = 12\,000$ kg; the maximum force you can apply is $3\,\text{t} = 29{,}420$ N because the 3 t rating means that it can hold up a 3 t mass which has a weight of 3000×9.8 N. So the maximum acceleration is $a_{max} = 29\,420/12\,000 = 2.45$ m s^{-2} $= 5.5$ mph s^{-1}. This means that if you speed up to 5.5 mph in 1 s, the rope will almost break. Of course, there will be friction and so to be safe I would recommend a factor of roughly two, an acceleration of about 3 mph s^{-1} would probably be safe. Here is an equation you can use (which does not include any safety factor): $a_{max} = 22(M_T/M_V)$ mph s^{-1} where M_V is the mass of the vehicle, and M_T is the mass rating of the towrope. If you are trying to tow up a hill which makes an angle θ with the horizontal, you need to apply a factor of $\sin\theta$ to the equation above, $a_{max} = 22(M_T/M_V)\sin\theta$, because some of the vehicle's weight is now directed down the hill instead of straight down. If the grade is 30^0, for example, $\sin\theta = \frac{1}{2}$.

The next question is a little different because rather than worrying about the tow line breaking you need to worry about traction of the wheels of the towing vehicle as well as its power, ability to deliver the energy.

A similar question: Watching a TV commercial showing how mighty a pickup truck is—it's towing the space shuttle, which weighs (according to the announcer) 292 000 lb (146 tons). Now I know that it's not as if the pickup is lifting 146 tons—I figure the load on the little pin hooking the shuttle to the pickup will be (initially) 146 tons times the coefficient of friction for the tarmac upon which both vehicles are riding—am I right?

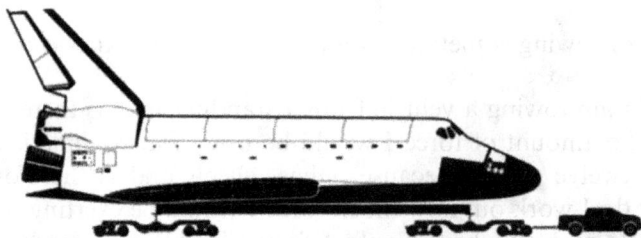

Answer: Assuming there is negligible friction in the bearings of the carriage for the shuttle, it is not hard to get the shuttle moving with a small acceleration. In the previous question, though, the numbers were much smaller than in your case where there is what appears to be a steel towing bar which would far exceed the strength of a towrope to tow things with weights of several tons rather than several hundred. So, with such a strong 'towrope' you might think that you could have as big an acceleration as you like. For example, if the breaking strength of the pin (probably the weakest link) were 100 tons, my little formula above would say that you could have an acceleration up to about $22 \times (100/146) \approx 15$ mph s^{-1} (0–60 in 4 s)! This will obviously not happen. There are two considerations you need to think about. First, the force which provides the acceleration is actually the static friction between the truck wheels and the road; the biggest this force can be is $f = \mu W$ where μ is the coefficient of static friction and W is the weight of the truck. For rubber on dry concrete, $\mu \approx 0.7$ and the weight of a Toyota Tundra pickup is about 3 tons, so $f = 3 \times 0.7 \approx 2$ tons; so, the maximum acceleration is only about 0.3 mph s^{-1}. The second consideration is how rapidly the truck can deliver the energy needed to move the load, in other words its power rating of about 300 hp. I calculate that the maximum acceleration with a 146 ton load would be about 4 mph s^{-1}. So, it appears that the main limiting factor on the acceleration is the possibility of the tires spinning. Keep in mind that these are all just rough estimates, but they give the general picture. (See appendix E.)

Many questions I have received ask things like if a box is full of birds which are flying, does it weigh less than if they were all roosting. These questions can get a bit convoluted although the question after the next one will be such a question. A simpler question is about a juggler.

Question: Is a juggler, while juggling three weights or any number really, lighter at all times than she would be if she merely carried the weight about her person? If so then by how much, when and why? If not then what does happen to their weight while they juggle at the various times they are and are not in contact with the juggled objects?

Answer: First, I am a stickler for the use of the word weight. The juggler's weight is the force which the Earth exerts on her and so it is always the same unless she overeats or goes on a diet. But, the apparent weight (what would be read by a scale she is standing on) depends what is going on with the balls. If all the balls are in the air at some time, her apparent weight will be her actual weight. If she is simply holding one ball, the scale will read her weight plus the ball's weight. If she is in the process of juggling one of the balls, she is exerting an upward force which will be larger than the weight of the ball (Newton's second law); but, because of Newton's third law, we can conclude that the ball exerts an equal and opposite force on her; and so the force read by the scale will be larger than the weight of the ball plus juggler. Only if she throws a ball downward would her apparent weight be smaller.

Now the bird-in-a-box problem.

Question: If there is a trailer full of birds and the birds are sitting on the bottom, does it weigh the same as if all the birds are flying?

Answer: (In this answer, when I say 'weigh' it means what a scale would read.) There is more than one answer to this question. Let us assume that the birds are hovering or moving with constant velocities. In that case, each bird stays in flight because the air exerts a force up on it equal to the bird's weight; but Newton's third law requires that the bird therefore exerts an equal downward force on the air. The air is part of the trailer, so the net weight of the whole truck is unchanged. Another possibility would be if the birds have an acceleration with a vertical component; the simplest example is that all the birds are in freefall inside (probably not what you had in mind by 'flying') in which case the birds would not contribute to the weight (neglecting any air friction or buoyancy) and the overall weight would be smaller. Or, if all the birds were at some instant accelerating upward, the air would be exerting an upward force on them greater than their weight so the trailer would measure heavier.

To finish off this section, here is a clever way to exert a force much bigger than the maximum force with which you can push.

Question: Today on NPR's *Cartalk*, someone called in a physics question. I would like to have a definite answer (very easy for you I'm sure). Here it is: A lady's car is stuck in the mud. She of course is alone with no phone and is a physicist. She ties a rope to her car bumper and a nearby tree. She then finds the mid-point of the rope and pushes with max effort which she estimates to be 300 N. The car just begins to budge with the rope at about a 5° angle. With what force is the rope pulling on the car? Ray, co-host of *Cartalk*, said to find the sine of 5 degrees and then multiply by 300. Then he changed it to cosine of 5 degrees and multiply by 300. If any of these is right, I don't understand why.

Answer: One of my favorite shows! Neither of the answers is right which is surprising since Tom and Ray are both are MIT grads. Here is how you do the problem (see figure 1.6). The point where she is pulling is in equilibrium, so the vector sum of the three shown vectors (her 300 N pull and the tensions in the two halves of the rope) must equal zero. The components perpendicular to her pull must add to zero, so the tension (T) in each side of the rope is the same. This comes from $T_1 \cos(5°) - T_2 \cos(5°) = 0$, so $T_1 = T_2 = T$. Similarly, the components parallel to her pull must sum to zero, so $300 - T\sin(5°) - T\sin(5°) = 0$. So, $T = 300/(2\sin(5°)) = 1721$ N.

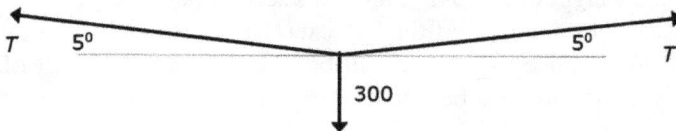

Figure 1.6.

1.6 Accelerated frames and fictitious forces

In section 1.1 we discussed noninertial frames, frames where Newton's first law is false and Newton's second law cannot be applied. A noninertial frame is, essentially, any frame which accelerates relative to any inertial frame. Amazingly, it is possible to force Newtonian mechanics to be valid in an accelerating frame if you judiciously add forces which do not exist, called *fictitious forces*. It is easiest to start the discussion of fictitious forces with a question involving linear acceleration.

Question: What is the force that causes you to fall over when a moving bus comes to an immediate stop? I'm having an argument with my teacher over what the answer is, it would be great if you could explain!

Answer: When the bus is stopping, it is accelerating and so it is a noninertial frame. That means that Newton's laws are not valid if you are riding inside the bus. But, if we watch you from the bus stop, Newton's laws do apply and we conclude that if you move with the bus, there must be a force which is causing you to accelerate also. Friction provides a force which, except under extreme circumstances, accelerates your feet along with the bus; but, unless you are holding on to something, there is nothing to provide a force on your upper body which therefore tends to keep going forward as the bus stops. All this says that the reason you fall forward is not due to any force, rather it is due to the lack of a force. There is, though, another way to look at this problem. If you are in an accelerating frame, like the bus, you can force Newton's laws to be true by adding fictitious forces. In the bus which has an acceleration a you can *invent* a fictitious force $F_{fictitious}$ on any mass m in the bus, $F_{fictitious} = -ma$; the negative sign means that the fictitious force points in the direction opposite the acceleration. If you do that, Newton's laws become true inside the bus and the force $F_{fictitious}$ may be thought of as being the force which provides your acceleration. Note that the acceleration is opposite the direction the bus is moving when it is stopping, and so the fictitious force is forward as you know if you have fallen over in a stopping bus. When the bus is speeding up you tend to fall backwards. Since there are two answers here, depending on how you choose to view the problem, maybe you and your teacher are both right!

From this you learn that the secret to making Newtonian mechanics work in noninertial frames is to add fictitious forces to masses m whose direction is opposite the direction of the acceleration a of the frame and of magnitude ma. Let us now re-examine the bicycle going around a curve which we looked at earlier. Here we encounter that best-known of nonexistent forces, the centrifugal force.

Question: Why when a cyclist is turning round a bend, does he lean inward?

Answer: Figure 1.7 shows the forces (real and fictitious) on the cyclist. The circle represents the center of mass of the system. Since he is moving in a circle of radius r and with speed v, he experiences a centripetal acceleration $a_c = v^2/r$ to the left. The forces on him are his own weight mg, the normal force N up from the road, and the frictional force F_f which is the force providing the acceleration. If you want to apply Newton's second law in the frame of reference of the cyclist, which is not an

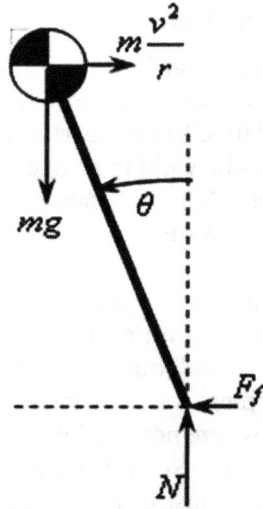

Figure 1.7. A bicycle turning a curve.

inertial frame, you must add the fictitious centrifugal force ma_c as shown in figure 1.7. Note that if he were not leaning, there would be an unbalanced torque about the point where the tire touches the ground, $\tau = mLv^2/r$ where L is the distance to the center of mass, which would cause him to rotate clockwise, that is to fall over. When he leans, though, the weight also exerts a torque, so the two torques can balance if the angle is just right: $mgL \sin \theta = mLv^2 \cos \theta/r$, or $\theta = \tan^{-1}(v^2/rg)$.

The centrifugal force is often suggested in sci-fi movies or books as a source of artificial gravity. Imagine that you are inside a very large hollow cylinder with radius R which is rotating around its axis such that the speed of the outer surface of the cylinder is v. You and the cylinder, your home, are in outer space with no gravity around. If you are at the inside surface and rotating with the cylinder, you are in a noninertial frame with acceleration, pointing toward the axis, of v^2/R. Then, if v and R have been chosen such that $v^2/R = g$, you will experience an apparent force mg, just as if you were standing on the surface of the Earth. Of course, R must be much larger than your height or else your head and feet will experience different accelerations. Suppose that $R = 200$ m; taking $g \approx 10$ m s^{-2}, $v = \sqrt{2000} = 45$ m s^{-1}. The circumference of the cylinder is $2\pi R = 1257$ m, so the time to make one revolution is $1257/45 = 280$ s $= 0.47$ h; the rotation is at a rather lazy rate of about two rotations per hour, a reasonable model for a space habitat. (See appendices C and D.)

The work–energy theorem, $W = \frac{1}{2}mv_{\text{final}}^2 - \frac{1}{2}mv_{\text{initial}}^2$ is derived (see appendix A) starting with Newton's laws. The following question asks whether this is true in noninertial frames.

Question: Is work–energy theorem valid in noninertial frames?
Answer: The work–energy theorem says that the change in kinetic energy of an object is equal to the work all forces do on it. Imagine that you are in an accelerating rocket ship in empty space, a noninertial frame. You have a ball in

your hand and you let go of it. You observe this ball to accelerate opposite the direction in which the ship is accelerating and therefore see its kinetic energy change. But, there are no forces acting on it so no work is done. Another way you could come to this conclusion is that the work–energy theorem is a result of Newton's laws and Newton's laws are not valid in noninertial frames. You can, though, force the work–energy theorem to be valid if you introduce fictitious forces, a way to force Newton's laws to work in noninertial frames. If you invent a force on the objects of mass m in the accelerating (a) rocket ship above of $F_{\text{fictitious}} = -ma$, this force will appear to do the work equal to the change in kinetic energy. (See appendix C.)

1.7 Wagers, arguments and disagreements

The Physicist is often called on to settle disputes. There are already a couple examples in earlier sections, the accelerated bus question and the badminton/tennis question. Here are a couple of others.

Question: My friend and I had a drunken argument. I would like independent council to weigh in (there's $300 on the line). I was given a unique bottle opener by a friend who is a brewer for a craft brewery in the northeast. It is a flat piece of wood with a smooth screw embedded near one end. The argument is as follows.

- Person A: There is less force required to open the bottle pulling up with the screw positioned between the cap and the user (top panel in figure 1.8).

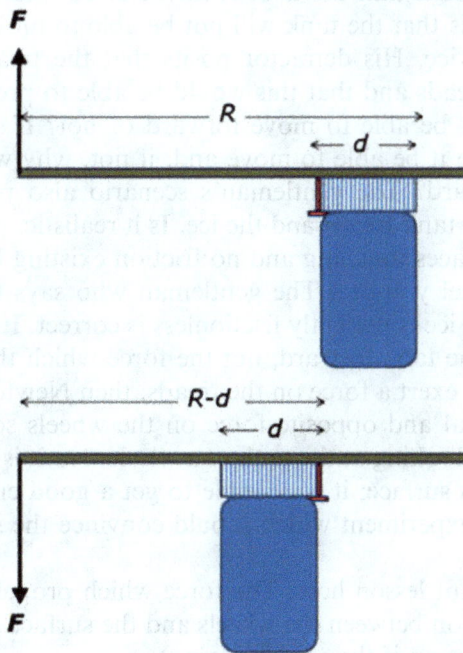

Figure 1.8. A unique bottle opener, person A above, person B below.

- Person B: There is less force required to open the bottle pressing down with the cap positioned between the screw and the user, (bottom panel in figure 1.8).

Can you prove either argument successfully?

Answer: The questioner also provided the information that $R = 94/16''$ and $d = 17/16''$. To answer the question I will compute the force which the nail exerts on the bottle top for equal forces by the user. Whichever of these is the biggest is the winner. Doing this is a simple first-semester physics statics problem, most easily done by summing the torques in each case about the point on the bottle cap just opposite the nail; that point is a distance R from the end where F is applied for person A and a distance $R - 2d$ for person B. I find that the nail exerts a force of $F_B = F[(R/d) - 2]$ for person B and a force of $F_A = F(R/d)$ for person A; person A is the winner of the bet. For your numbers, $R/d = 5.53$ and the ratio of the forces is $F_A/F_B = 5.53/3.53 = 1.57$, making option A 57% bigger, quite definitive. (If a \$300 bet is really on the line, don't forget to reward *The Physicist*!) (See appendix A.)

I am pleased to report that these barroom physics enthusiasts did indeed send a generous contribution to *Ask The Physicist*! Here is another question, this one about friction.

Question: I am writing in the interest of hopefully resolving a question which had arisen in my workplace. One gentleman poses the hypothetical situation of a motionless tank sitting on solid ice which he describes as 'very slick and smooth—so much so that if one were to toss a penny across the surface then it would glide on endlessly'. He posits that the tank is then started and attempts to move forward. His position is that the tank will not be able to move as the treads would simply spin on the ice. His detractor posits that the treads are moved by the wheels inside the treads and that this would be able to propel the tank forward. So, would this tank be able to move forward or not? If so, what properties of physics would make it be able to move and, if not, why would this tank not be able to move forward? The gentleman's scenario also posits that there is no friction between the tank treads and the ice. Is it realistic, physically speaking, to posit these two surfaces touching and no friction existing between them?

Answer: How genteel you are! The gentleman who says that the tank will not move forward if the ice is perfectly frictionless is correct. It is the force of friction which accelerates the tank forward, not the force which the wheels exert on the treads; if the wheels exert a force on the treads, then Newton's third law says the treads exert an equal and opposite force on the wheels so the two cancel each other out if you are looking at the tank as a whole. No it is not possible to have a perfectly frictionless surface; it is possible to get a good enough approximation, however, to do an experiment which should convince the second gentleman.

There is an important lesson here. The force which propels a wheeled vehicle is the force of static friction between the wheels and the surface they are rolling on (or the force of kinetic friction if they are spinning).

Here is a dispute regarding torques and center of gravity.

Question: I have a question related to weight/mass placement on a bar. My friend and I are weight lifters. We got into a discussion about the center of gravity on the bar. Here is the question. We are using a 45 lb plate on each side and also have a 5 and 10 on each side, each taking up the same space and the end of the bar is the same distance from the last weight and will not change. Does it change anything if the weights are not in the same order, from one side to the other? My friend says the side with the 45 lb plate close to the end is slightly heavier because the ratio has changed. I say nothing has changed because the weights on the bar are still taking up the same space. I believe it would only change if the distance to the end of the bar is changed, which it is not. I hope I explained this well enough.
Answer: Assuming that the bar itself is uniform (has its center of gravity (COG) at its geometrical center), the COG of the total barbell depends on the location of the weights. Relative to the center of the bar, the position of the center of gravity may be written as $\text{COG} = (45x_1 + 10x_2 + 5x_3 - 45x_4 - 10x_5 - 5x_6)/120$ where the x_i are the distances of weights from the center. Suppose that the weights are placed symmetrically ($x_1 = x_4$, $x_2 = x_5$, $x_3 = x_6$); then $\text{COG} = 0$, the center of the bar. Now, suppose we interchange two of the weights, exchange the 45 lb with the 10 lb on one side:

$$\text{COG} = (45x_2 + 10x_1 + 5x_3 - 45x_4 - 10x_5 - 5x_6)/120$$
$$= (45x_1 + 10x_2 - 45x_2 - 10x_1)/120$$
$$= (35/120)(x_1 - x_2);$$

since $x_1 \neq x_2$, $\text{COG} \neq 0$, the barbell is no longer balanced. If that explanation is too mathematical for you, try a more qualitative argument. Each weight W a distance D from the center exerts a torque about the center and the magnitude of that torque is WD. The net torque due to all weights must be zero if the bar is to balance at its center. This means that the sum of all the WDs on one side must be precisely equal to those on the other if the barbell is to be balanced about its center. If you change the D on only one side, the bar will not be balanced at its center. (This qualitative argument is just the mathematical argument in words.) What certainly does not change is the total weight. (See appendix C.)

1.8 These are a few of my favorite things

This final section in chapter 1 collects some of my favorite questions and answers, questions which I found very interesting or questions from which I learned or just plain cool questions.

This first question surprised me greatly. I would never have guessed that gravity would pull two dice separated by a few centimeters together in a matter of hours.

Question: Two dice are suspended in outer space with no visible forces acting on them. Their centers of mass are 10 cm apart, and they each have an identical mass of .0033 kg. How long would it take for the force of gravity between them to cause them to touch? (We will assume they are volumeless for ease in calculation).

Answer: This seems a very difficult problem because the gravitational force between them changes as they get closer and so it is not a case of uniform acceleration. However, this is really just a special case of the Kepler problem (the paths of particles experiencing $1/r^2$ forces) which I have done in detail before. You can go over that in detail. For your case, $K = Gm_1m_2 = 6.67 \times 10^{-11} \times (3.3 \times 10^{-3})^2 = 7.26 \times 10^{-16}$ N·m^2 kg^{-2}, the reduced mass is $\mu = m_1m_2/(m_1+m_2) = 0.0033/2 = 1.65 \times 10^{-3}$ kg, and the semimajor axis $a = 2.5$ cm $= 2.5 \times 10^{-2}$ m. Now, from the earlier answer, $T = \sqrt{(4\pi\mu a^3/K)} = 5.98 \times 10^4$ s. The time you want is $T/2 = 2.99 \times 10^4$ s. This is only 8.3 h and seemed too short to me. To check if the time is reasonable, I calculated the starting acceleration and assumed that the acceleration was constant and each die had to go 5 cm; this time should be longer than the correct time because the acceleration increases as the masses get closer. The force on each die at the beginning is $K/r^2 = 7.26 \times 10^{-16}/0.05^2 = 3.04 \times 10^{-13}$ N; so, the resulting initial acceleration is $F/m = 3.04 \times 10^{-13}/3.3 \times 10^{-3} = 9.21 \times 10^{-11}$ m s^{-2}. So, assuming uniform acceleration, $0.05 = \frac{1}{2}at^2 = 4.61 \times 10^{-11}t^2$. Solving, $t = 3.3 \times 10^4$ s. So, the answer above is, indeed, reasonable.

The following question is one of my very favorites because it is so deceptively simple and yet so subtle to understand. I pondered this on and off for several days and finally needed to talk it over with some other physicists. I would like to acknowledge a very useful discussion over pizza with friends and colleagues A K Edwards, W G Love, R S Meltzer and R L Anderson.

Question: My question has to do with traction and the movement of a wheel (a wheel alone). Traction is essential for its movement both linear and circular. But if we throw a wheel forward it rolls some meters and then it stops (and falls). Which force is responsible for the decrease in its velocity? Because if traction is parallel to the ground facing backwards, then linear movement's negative acceleration is explained but not angular negative acceleration. If traction is parallel to the ground facing forward then angular negative acceleration is explained but not linear. If traction is zero then which force decreases both velocities linear and angular?

Answer: One of the reasons I love doing *Ask the Physicist* is because I often learn things I did not know or had never thought about. You would think that a guy who has been teaching introductory physics courses for nearly 50 years would find this question simple. But, indeed I was puzzled by it because, as I have found by thinking about it and talking to some friends, I wasn't thinking beyond the friction force (which questioner calls traction) being simply the only force in the horizontal direction and obviously stopping the forward motion after some distance. I never addressed the angular acceleration of the wheel before. What frictional forces are important to understand the rolling of a wheel? Most introductory physics classes talk only about the contact forces of static friction and kinetic friction. Kinetic friction is not applicable to this problem because the wheel is not slipping on the ground, and static friction might be important, but not necessarily. If we have a round wheel rolling on a flat horizontal surface (don't look at figure 1.9 yet!), there are three possible forces—the weight which must be vertical, pass through the center of mass, and (assuming it is a uniform wheel) pass through the point of contact; the friction, which must be parallel to the surface and pass through the point of contact; and the normal force which must be perpendicular to the surface and pass through the point of contact. If you now sum torques about the point of contact (as noted by the questioner), there are none! So, there can be no angular acceleration; if we have stipulated that the wheel does not slip, then there can be no linear acceleration either and the wheel will roll forever and no friction is required. But we all know better! A real wheel will eventually slow down. The key is that there is no such thing as a perfectly round wheel or a perfectly flat surface, one or both must be deformed. In that case, we have to think about a new kind of friction called rolling friction, the friction the wheel has because of the rolling. This is different from the static friction, and static friction may still be present to keep the wheel from slipping. A perfectly round wheel cannot have rolling friction as I showed above, it must deform which means that there is no longer a 'point (or line) of contact' but now an area of contact. Since the normal force is only constrained to act somewhere where the two are in contact, it is now possible (in fact inevitable) that this force will not act through the center of mass of the wheel. That is the whole key to answering this question. So, finally, the answer: refer to figure 1.9 where I have drawn the forces *mg*, *N*, and *f*. The weight is still constrained to be vertically down and pass through the center of mass (blue cross). The normal force is constrained to be vertical and act somewhere where the wheel and ground are in contact, drawn a distance *d* to the left. The frictional force (which now includes both static and rolling friction) is constrained to act at the surface and parallel to it. I choose a coordinate system with *x* to the left and *y* up; the axis (red cross) about which I will sum torques is at the ground directly under the center of mass and positive torque results in an angular acceleration which is positive when acceleration of the center of mass is positive (counterclockwise around the axis). All is now straightforward: $\Sigma F_x = -f = ma$, $\Sigma F_y = N - mg = 0$, $\Sigma \tau_x = Nd = I\alpha = Ia/L$ where I is the moment of inertia about the red cross and L is the distance from the red cross to the blue cross. Finally, $N = mg$, $a = -f/m$, and $d = fI/(Lm^2g)$. (See appendix C.)

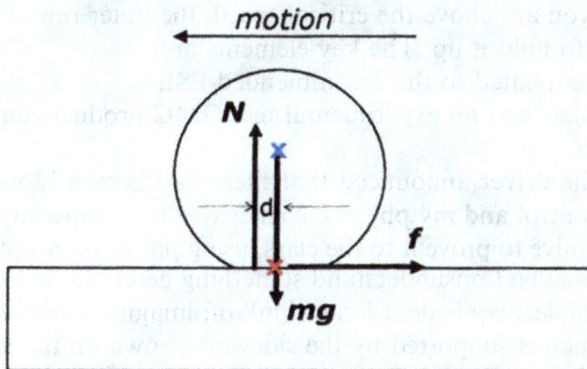

Figure 1.9. A 'real' wheel rolling.

Sometimes I find myself in a debate which, like in the following case, lets me view a question I had dismissed with an open mind and coming around to common ground with the questioner. Here is the disturbing thing, though: after finally having given an answer to his question which explained how a blown-out tire at high speed could retain its roundness (but definitely not without the presence of the sidewalls), the questioner wrote thanking me for proving that sidewalls were not needed at high speeds! Guess he was less able to listen to the other side of the argument than I was.

Question: In 1973 a physics instructor explained that the sidewalls of a regulation tire need not be present if the velocity of the vehicle was above a speed of 65+ mph. I tried to explain this to family members at Christmas and was scoffed at and then ridiculed. The physics instructor had previously worked at a GMC/Chevrolet plant. His job had been to change out instruments on GM cars running around a track and in excess of 100 mph. In one of the test runs his driver advised him that they had had a blowout and he needed to get out from under the dash quickly and get safety belted in. Then the driver slowed down and at some critical speed he almost, but not quite, lost control and they did not crash but came close to it. The instructor was a good instructor in that he made the physics relevant to the real world. Also this is why tires need sidewalls as they won't hold up in gravity and below a specific velocity.

Answer: This is nonsense. If there is no air pressure to connect the tire to the axle, which would be the case if there were no sidewalls, what is going to hold up the weight of the car?

Follow-up question: No not really. If you get the tire up to speed, as well as providing forward momentum, the circumference and the center point about which the tire is rotating will hold the tire up even if there is a blow out as the forward speed or acceleration is sufficient to hold it up and will prevent collapse of the tire above a critical speed. Once the speed drops below this critical speed, the tire will start to collapse and, according to the physics instructor, all hell broke loose on the track and only the driver's expertise insured that they were able to stop safely. If you are above the critical speed, the outer rim of the tire need not have anything to hold it up. The key elements are:

1. Tires are inflated to the recommended PSI.
2. The vehicle was an experimental test GMC product running in excess of 100 mph.
3. When the driver announced that there had been a blowout, the car was under control and my physics teacher was not aware anything was amiss.
4. He was alive to prove it to the class, using physics concepts that escape me.

Answer: Well, maybe I misunderstand something here, but let's boil this problem down to the simplest equivalent I can think of: imagine a tire with sidewalls and just an axle which is supported by the sidewall, shown on the left in figure 1.10. Now, we would agree, I believe, that if the sidewall suddenly disappeared, the axle would fall because there would be nothing holding up that weight. How is that

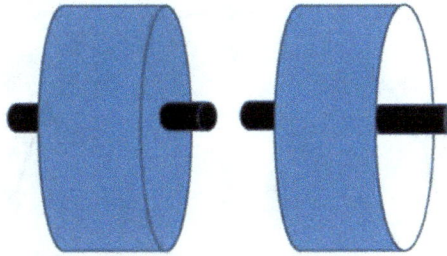

Figure 1.10. Tires with (left) and without (right) sidewalls.

situation any different if the car is moving? So, let's agree that 'the sidewalls… need not be present' is wrong because there has to be some physical contact of the outer surface of the tire and the axle. So, my first answer was a knee-jerk response to the notion that the sidewalls were not needed.

The answer you will like: However, there is still a way that you might have a point. When the blowout occurs, the pressure inside the tire is lost; this pressure is typically $30 \ \mathrm{PSI} = 21\,000 \ \mathrm{N\,m^{-2}}$ above atmospheric pressure (which is about $100\,000 \ \mathrm{N\,m^{-2}}$). If the car is sitting still, this loss of pressure results in the wheel collapsing because the sidewalls alone are insufficient to hold up the weight of the car unless the force due to the pressure pushing on the outer part of the tire holds the sidewalls taut. Now, imagine that you are driving with some speed V and viewing a spinning tire from its axis, you see every point on the outer surface of the wheel accelerating with an acceleration V^2/R where R is the radius of the tire. Therefore, every little piece of the tire with mass m experiences a (fictitious) force (called the centrifugal force) of mV^2/R. That would be equivalent to there being a pressure P exerted on that little piece of tire of $P = mV^2/(aR)$ where a is the area of that little piece. But, every little piece behaves like this, so it is equivalent to a pressure of $P = MV^2/(AR)$ acting on the outer surface where M is the mass of the tire (assuming the sidewalls are a small fraction) and $A = 2\pi RW$ is the area of the outer surface and W is the tread width. So, if that pressure is equal to $21\,000 \ \mathrm{N\,m^{-2}}$, it will be like the blowout never happened! I took $R \approx 16 \ \mathrm{in} \approx 0.4 \ \mathrm{m}$, $W \approx 12 \ \mathrm{in} \approx 0.3 \ \mathrm{m}$, and $M \approx 20 \ \mathrm{lb} \approx 9 \ \mathrm{kg}$ and solved $21\,000 = MV^2/(AR) = MV^2/(2\pi R^2 W)$ and found $V = 27 \ \mathrm{m\,s^{-1}} = 60 \ \mathrm{mph}$. (Incidentally, the 'forward momentum' has nothing to do with it.)

The following question was fun because it got me thinking about how strong a rotating structure has to be. I am not an engineer and only claim to do an order-of-magnitude calculation here, but it seems to jive pretty well with estimates given by the questioner.

Question: I've read about space habitat concepts for a while and I've run into an interesting concept. The concept I've run into is the McKendree cylinder which is basically an O'Neill cylinder made of carbon nanotubes. The O'Neill cylinder made of steel would be 32 km long and 6 km in diameter. The McKendree

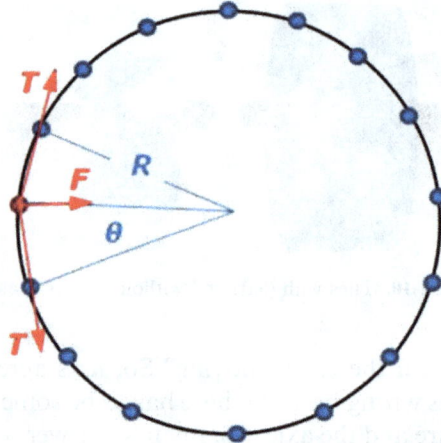

Figure 1.11. Forces on the edge of a rotating space station.

cylinder would be 4600 km long and 460 km in diameter. And the maximum length for the McKendree cylinder is 10 000 km and diameter of 1000 km. So the McKendree could be built a lot bigger than an O'Neill one because the carbon nanotubes have greater endurance. But a habitat of thousands of kilometers seems to be really big when compared to what we can build from other materials. And as I recall we don't have any ways to produce carbon nanotubes in large quantities. Is it theoretically possible to build a habitat 10 000 km long and 1000 km wide out of carbon nanotubes? And is the McKendree cylinder more of a theoretical design than a practical design that actually could be built?

Answer: I presume that the issue is more a strength issue than anything else. To illustrate how the strength of the material and its mass determine the size the habitat can be, consider a rotating string of beads, each of mass m. The rotation rate must be such that $a = v^2/R = g$ where v is the tangential speed of each bead. Therefore each bead must experience a force $F = mg$. This force can only come from the two strings attaching each bead to its nearest neighbors and, from figure 1.11, $F = mg = 2T \sin \theta$. But, we will imagine many, many beads on this string and we will call the distance between them d; so we can make the small angle approximation that $\sin \theta \approx \theta = d/R$. Solving for T, $T = mgR/(2d)$. Now imagine that the beads are atoms; d will be about the same for steel or carbon, g is just a constant, $m_{steel} \approx 5 m_{carbon}$, and the Young's modulus of carbon nanotubes is about five times bigger than steel, $T_{steel} \approx T_{carbon}/5$. So, $R_{carbon}/R_{steel} \approx (T_{carbon}/T_{steel})/(m_{carbon}/m_{steel}) \approx 25$. Your numbers are $R_{carbon}/R_{steel} \approx 460/6 = 77$; I would have to say that my calculation is pretty good given that I have made very rough estimates and I am not an engineer! I do not know what considerations would limit the length of the habitat. (Of course, neither of these models is currently practical to actually build, so call them theoretical if you like. However, there would certainly be no problem building them if resources and manufacturing capabilities were available.) (See appendix D.)

From Newton to Einstein
Ask the physicist about mechanics and relativity
F Todd Baker

Chapter 2

Special relativity

2.1 Overview

Open just about any textbook which includes the theory of special relativity and you will find that the first thing covered will be the Michelson–Morley experiment and the failure to detect the luminiferous æther. Every wave that physics knew traveled through some medium—sound through air, water waves through water, standing waves through a violin string—and if you took away the medium, the wave no longer could exist. Light was a puzzle, though, since it seemed to travel through a vacuum, empty space. It was therefore assumed that there must be some medium, the luminiferous æther, which supported light waves and permeated all of the Universe. Generations of physicists and physics students have grown up believing that the failure to detect any evidence of the luminiferous æther was the impetus for the theory of special relativity. In fact, Albert Einstein in his 1905 paper, 'On the Electrodynamics of Moving Bodies' was motivated by a longstanding curiosity about the nature of electromagnetism and the properties of light. Although he was apparently aware of the failure to detect the æther, since it was mentioned briefly in the introduction to the paper, he said in later life that he did not really remember that it played any role in his development of relativity.

Although it is not a focus of this book, to understand the birth of the theory of relativity one must have at least a qualitative understanding of electromagnetism. During the 18th and 19th centuries many scientists had studied the properties of electric and magnetic forces as well as the relations between the two. Around 1870 James Clerk Maxwell took all that was known about electromagnetism and condensed it into his four famous equations. A summary, in words, of the essence of Maxwell's equations can be found in one of my *Ask The Physicist* answers:

Answer: The laws of electromagnetism are perfectly symmetric: a changing magnetic field causes an electric field and a changing electric field causes a magnetic field. The first of these is called Faraday's law and the second is part of Ampere's law. You seem to think that only a permanent magnet is magnetism. In fact, any moving electric charge causes a magnetic field. The most common source of magnetic fields is simply an electric current. Here are some facts about electric and magnetic fields:

- electric charges cause electric fields,
- electric currents (moving charges) cause magnetic fields,
- changing magnetic fields cause electric fields, and
- changing electric fields cause magnetic fields.

An amazing result derived from Maxwell's equations was that they predicted the existence of waves of electric and magnetic fields. The speed c of these waves was found to be determined by only two well-known constants, the permittivity of free space ε_0 and the permeability of free space μ_0, $c = 1/\sqrt{(\varepsilon_0 \mu_0)} = 3 \times 10^8 \, \text{m s}^{-1}$. (Essentially, these constants quantify the strength of electric and magnetic fields, see appendix F.) Surely it is not just a coincidence that this is precisely the speed of light, well known and accurately measured by the 19th century. Although it had been long known that light was a wave, it was not known what was doing the 'waving'; now it was clear that electric and magnetic fields comprised light waves.

Presumably, Maxwell's equations were for a frame of reference at rest in the æther. To find the form in another inertial frame which has a velocity v in the $+x$ direction should have been simple—just transform the equations by replacing x by $x - vt$ everywhere; this is called a Galilean transformation (see appendix G). The trouble is that, in doing that, the transformed equations contained impossible contradictions. The Dutch physicist Henrik Lorentz found a different transformation which seemed to work but there was no basis for it, it was purely empirical. Einstein, in his 1905 paper, actually showed that Lorentz's transformation resulted from applying two simple postulates:

1. The laws of physics are the same in all inertial frames of reference; this is called the principle of relativity.
2. The speed of light is the same for all observers regardless of their motions or the motion of the source.

In fact, only the first postulate is needed because Maxwell's equations are laws of physics and they predict that the speed of light in vacuum depends only on known constants of nature, not which inertial frame you happen to be in.

In the more than a decade of answering questions I have never written down the Lorentz transformation and I do not intend to start now! (I do, however, write them in appendix G for completeness.) Once you have accepted that the speed of light is a universal constant, it follows rather nonmathematically that moving clocks run slow, moving sticks are shorter, moving masses get bigger, and all the other fascinating results of relativity which have so changed the way we think about how the Universe works.

2.2 Newtonian mechanics is wrong

Electromagnetism is the key that opened the door to special relativity, but relativity is mainly a new version of Newtonian mechanics. But that can only mean that the old mechanics is wrong. But how can that be? For more than two centuries Newtonian mechanics had been perfected into a beautiful science describing virtually everything around us. As we shall see, the effects on mechanics of the speed of light being universal only become apparent when objects have speeds which are comparable to the speed of light; of course, our everyday life never encounters such speedy objects, so whatever we come up with as a new theory must reduce to Newtonian mechanics if the speed v is small compared to c, $v \ll c$.

The notion that the speed of a ray of light is the same in all inertial frames flies in the face of our intuition. Consider the following question.

Question: What is the velocity of a person if he is on a train that is traveling at $50 \, \text{m s}^{-1}$ east and he is running at $2 \, \text{m s}^{-1}$ west?
Answer: Every time I get this kind of question (which is often) I have to emphasize— who is measuring the velocity? If it is somebody in the train, the velocity is $2 \, \text{m s}^{-1}$ west. If it is somebody on the ground, it is $48 \, \text{m s}^{-1}$ east. (See appendix G.)

This is called velocity addition. It is common sense, right? If you walk $2 \, \text{m s}^{-1}$ with the train your speed as observed by somebody at the station is $52 \, \text{m s}^{-1}$, backwards on the train at $2 \, \text{m s}^{-1}$ would be $48 \, \text{m s}^{-1}$. The velocity addition formula in one dimension may be written as $v' = v + u$ which means, in words, that if you are traveling with speed u and there is another car approaching you with speed v, his speed of approach to you is v'. For example, if you are going 60 mph north and another car is going 50 mph south, you see the other car approach you with a velocity of 110 mph: $v' = 50 + 60 = 110$ mph.

But, suppose we apply this to light with speed c in our frame. In another frame moving toward you with half the speed of light and opposite the light direction, the other frame should see the light to have a speed of $1.5c$. but, it does not, it also sees c. So, if the velocity addition formula is wrong, it should show up at high speeds. Here is a question which illustrates this problem.

Question: Hello, I used to have a Feynman book that had this scenario and I forgot how he explained it. I have since lost the book and was wondering if you could explain it. I have a spaceship moving at $180\,000 \, \text{km s}^{-1}$ and inside of that spaceship I have another spaceship moving at $180\,000 \, \text{km s}^{-1}$. To the observer on the ground the second spaceship is moving at $360\,000 \, \text{km s}^{-1}$. That exceeds the speed of light please explain what would happen.
My first response: Your recollection is wrong. I am sure Feynman never said the speed of the second space ship exceeds the speed of light because it doesn't.
Follow-up: Ya, he may have never said this, but can you explain what would happen in that scenario?
Answer: The equation which describes what is called 'velocity addition' in relativity is $v' = (u + v)/[1 + (uv/c^2)]$ where u is the speed of the first ship, v the speed of the second ship (relative to the first), c the speed of light ($300\,000 \, \text{km s}^{-1}$), and v' is the speed of the second ship seen by a stationary observer. Note that if u and v are both very small compared to the speed of light, then the quantity (uv/c^2) is very close to zero so that $v' \approx (u + v)$, which is what you expect to be correct, is approximately true. However, in the example you cited the speeds are not

small compared to c (they are 60% of c). If you do the arithmetic you will find that $v' = 265,000$ km s^{-1}.

So, why does your intuition fail here? Quite simply because intuition is based on experience and you have absolutely no experience with objects moving with half or 60% the speed of light. The principle of relativity demands that c be a universal constant and that means that Newtonian mechanics must be wrong. Strictly speaking, the velocity addition formula is Galilean relativity (see appendix G), it is not part of what Newton introduced as mechanics. Still, if it is wrong, the implication is that Newtonian mechanics is wrong for the following reason. In Galilean relativity the addition formula for acceleration addition is $a' = a$, that is all inertial frames see the same acceleration. If this were not true, then if you exerted a force on a mass m, one frame would say that the force was ma while the other would say the force exerted was ma'. How could they disagree on the magnitude of a force?

In the following sections you will see many unexpected and astonishing consequences of the speed of light being a universal constant of nature.

2.3 Relativity of time, time dilation

The thing I find is most surprising to those first encountering special relativity is that all clocks in the Universe do not tick at the same rate. It never occurs to us that a clock moving by might be running slowly, not because it is a faulty clock and not because it 'looks like' it is running slower. Time really runs slower in a moving frame. The reason is that when Galilean relativity (see appnedix G) is revised to allow for the constancy of the speed of light, it is found that time and space are irrevocably tied together. In the one-dimensional case, where the relative velocity is along the x directions, not only does x' depend on v and t, t' depends on v and x. The Lorentz transformation, shown in appendix G, demonstrates this. However, trying to understand everything starting with the Lorentz transformation just adds mathematical opacity for the average layperson. In more than a decade of answering questions I have never found it necessary to write it down. There are other, more qualitative, ways to 'skin this cat'. But, you do need to accept that c is a universal constant. Since time is the intuitive hangup, I start there. Imagine a clock sitting on your desk which consists of a light source and a light detector at one end and a mirror at the other, the source aimed at the mirror; the two are separated by a distance L. The source emits a short pulse of light, the light goes up to the mirror in the time L/c, the reflected light ruturns to the detector in time L/c, another pulse is sent and the clock goes 'tick'. So, the time between ticks is $\tau = 2L/c$. For example, if $L = 1.5$ m, the time between ticks is $2 \times 1.5/3 \times 10^8 = 10^{-8}$ s. So this clock ticks every 10 ns. This is a perfectly good clock and any clock you have sitting next to it will certainly run at the same rate—when this clock ticks off 10 years, so will a pendulum clock, so will an atomic clock, so will your wrist watch; even a biological clock like your body will tick off 10 years. Now, your friend, who also thinks this is a great clock, is moving past you with some very large speed. When you watch her clock,

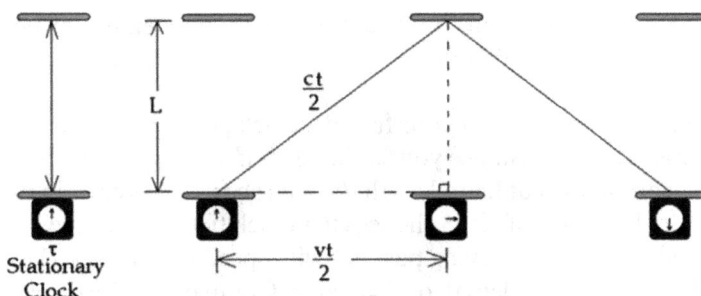

Figure 2.1. Stationary (left) and moving (right) clocks.

you will see that the path followed by the light is no longer of length L each way but a longer distance (see figure 2.1) given by $\sqrt{[L^2 + (vt/2)^2]} = (ct/2)$ where t is the time you see her clock tick once. So, solving this, you find that $t = (2L/c)/\sqrt{[1 - (v/c)^2]} = \tau/\sqrt{[1 - (v/c)^2]}$. Her clock, which we have agreed is a perfectly good clock, runs more slowly than yours by a factor of $\gamma = 1/\sqrt{[1 - (v/c)^2]}$ (called the gamma factor). This is called *time dilation*. For example, if the speed of the moving clock is 80% the speed of light, $\gamma = 1/\sqrt{[1 - (0.8)^2]} = 1/0.6 = 1.67$, so when 1.67 s ticks on your clock, only 1 s ticks on hers.

2.4 Relativity of length, length contraction

Having just studied the light clock, the following question provides a perfect introduction to length contraction, how moving lengths are actually shorter.

Question: I was wondering if you could explain how length contraction works. I've already done some background research and I understand the mathematical reasons my textbook gives me, I was just wondering if you could give some kind of analogy that would enable me to picture the effects of length contraction, and better yet allow me to explain it to my friends in a way they can understand.

Answer: I don't know a simple one-step way to intuitively understand length contraction. But, I know a good two-step way starting with intuitively understanding time dilation. Time dilation is pretty easy to understand in one simple example, the light clock. From that point, length contraction can be understood as a natural consequence of time dilation. Here is how it goes. Imagine a bomb which has a fuse of 1 s. A bomb is a clock which ticks once. If that bomb is moving by us with a speed of 99% the speed of light, the elapsed time before detonation measured by an observer at rest will be $1/\sqrt{[1 - (0.99)^2]} = 7.09$ s. So, the distance it will travel is $0.99 \times 3 \times 10^8 \times 7.09 = 2.1 \times 10^9$ m. But, the bomb will measure in his own frame that he should last 1 s and go $0.99 \times 3 \times 10^8 \times 1 = 2.97 \times 10^8$ m, only about 1/7 the distance we measure. If you think about the distance we measure as a long stick of length 2.1×10^9 m in its own frame, then the bomb sees this stick moving by him with speed 99% the speed of light, so to reconcile his results with ours he must measure that length to be $2.1 \times 10^9 \times \sqrt{[1 - (0.99)^2]} = 2.97 \times 10^8$ m.

So, the idea is that if you watch your friend's clock go past you, the elapsed time for one of her ticks is $\gamma\tau$, so the distance you see her go is $d = v\gamma\tau = 2v\gamma L/c$. Imagine that you have a stick on your floor that has a length d which she just traverses in one click of her clock. Now, from her point of view, she sees her clock tick with a time $2L/c$ and in that time the stick, which she sees moving past her with speed v, just passes her. That means that she sees the stick having a length of $d' = 2vL/c$. Comparing, $d' = d/\gamma = d\sqrt{[1 - (v/c)^2]}$, a shorter length. This is called length contraction. It is important that length contraction occurs only along the direction of motion, not perpendicular to that direction.

2.5 The twin paradox

Perhaps the most famous of relativity examples is the twin paradox where one brother travels to a distant star and back and the other stays home on the Earth. The following question illustrates why it is called a paradox.

Question: In the 'twin paradox', the twin that 'moves' ages more slowly. But don't they both 'move', since movement is relative? So, using the same logic, wouldn't the twin at home age less than the twin in the spaceship since the twin at home (relatively) 'moves' away from the spaceship? How can time be slower for one than the other since they both move, relative to each other?

Answer: You have hit on why they call it a paradox! However, there is really no paradox at all because there is an inherent asymmetry between the two twins. Think of the 'distance' to the destination star as a stick between the Earth and the star. The moving twin sees this stick contracted because of his motion whereas the earthbound twin does not. Hence, the moving twin sees a shorter distance he must travel and so it takes him less than the (classically) expected time. The following is the answer to an earlier question about the twin paradox.

Answer: My favorite way to understand the twin paradox is to suppose that each twin, using his own clock, sends a light pulse to his brother once a year. Each brother receives all the pulses from the other but the moving brother sends fewer since, because of length contraction of the distance to the object he travels to, he has less far to travel in his frame than his brother observes. To make things concrete and the arithmetic neat, I have chosen the distance out to which the traveling brother goes to be 8 light years away and back and his speed to be 80% the speed of light. This is shown in figure 2.2. Since the traveling brother sees, because of length contraction, a distance to the star of only 4.8 light years, he sends out six light pulses on the way out and six on the way back, and all are received by the earthbound brother, so both agree that he has aged 12 years. The earthbound brother, of course, sends out 20 pulses and the traveling brother receives them all so both agree that he has aged 20 years. So, there is no paradox because each brother agrees on the ages of both.

The twin paradox leads very naturally to the next section.

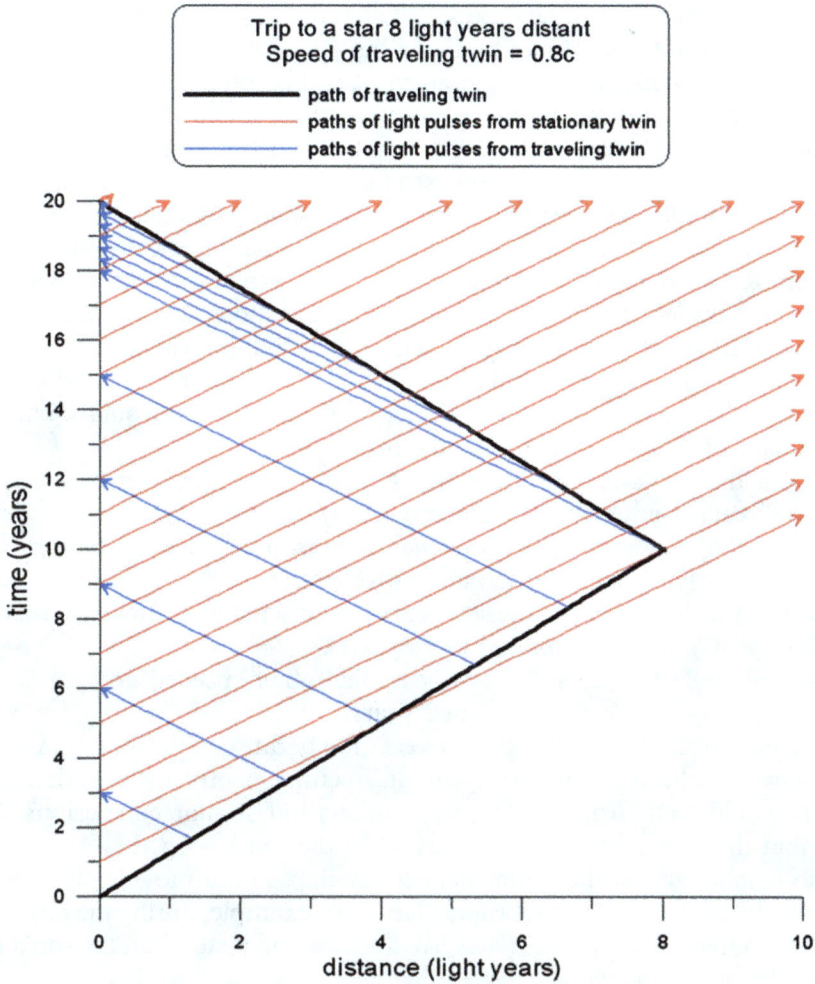

Figure 2.2. The twin paradox.

2.6 How things *look*, how things *are*

You will often read, even in textbooks, that length contraction and time dilation are how things appear to be. This is really misleading because it implies that we are dealing with optical illusions. In fact, how things appear to be are often quite different from how they actually are. Moving sticks can look longer than they are and moving clocks can look like they are running faster. Let us return to figure 2.2; during the first 18 years the earthbound brother sees his brother's clock to be going more slowly than his and during the last 2 years he sees it going much faster. And neither of these rates can be computed using the appropriate gamma factor; these time rates are not *how time is*, but rather *how time appears to be*. The following question gets at this time appearance issue.

Question: Say there is a father and son. The father is the first astronaut to attempt to 'time travel' into the future by flying his rocket around a black hole at extremely high velocities, so that for a given period of time (say an hour) it is a lot longer on Earth (say 10 h). My true question is, if the father could somehow communicate to his son in real time, what would this conversation sound like? Would the son hear his dad talking extremely fast? or would the distance between them make up the distance in 'time travel'. If it was the distance that is the factor, what if the dad was on a super train traveling at similar speeds on Earth?

Answer: Refer to figure 2.2 in section 2.5. As explained in the twin paradox answer, the traveling twin (the father in your question) takes 6 years to go each way while the stationary twin (son) has 10 years elapse each way. Each (father and son in your case) sends out one light pulse each year and by looking at the spacing of those pulses you can deduce how a conversation would sound. Here is a summary of how each sounds to the other:

- On the trip out, the father hears the son slowed down by a factor of 3 (2 yearly signals from home in 6 years).
- On the trip home, the father hears the son speeded up by a factor of 3 (18 yearly signals from home in 6 years).
- For the first 18 years, the son hears the father slowed down by a factor of 3 (6 yearly signals from dad in 18 years).
- For the last 2 years, the son hears the father speeded up by a factor of 3 (6 yearly signals from dad in 2 years).

Of course, it cannot really be a conversation because of the long transit times of the signals; rather each is just speaking, reciting poetry or something. Higher speeds would lead to more extreme numbers but similar conclusions. Overall, note that the father has aged 12 years while the son has aged 20 years. I always like to emphasize that how time *appears* to elapse on a moving clock is not the same as the time which actually *does* elapse; for example, during the last two years for the son, the father's clock *looks like* it is running faster than the son's whereas it *actually is* running slower.

Just as the rate at which a clock runs and the rate at which it appears to run are different things, the length objects are and the length they appear to be are not necessarily the same either, as the following question illustrates.

Question: Due to length contraction, you notice that a passing train appears to be shorter than when it is stationary. What do the people in the train observe about you? If you are on a train that is going really fast, do the people on the ground look shorter, longer, or the same?

Answer: Length contraction causes the lengths parallel to the direction of motion to be shortened. So, a fat man standing at the station would become a skinny man (side to side, but not front to back) and no shorter as measured by someone on the train. Similarly, as measured by someone on the platform, a fat lady standing on the train would become a skinny woman but no shorter. You will note that I did not say that these folks 'look skinnier' because physicists normally do not care how something *looks*, they care about how something *is*. This is a very important distinction and one which even authors of physics books often fail to make. How something looks may be very different from how something is. Hence, your question is incorrectly stated ('...appears to be shorter...') although I believe I know what you meant. Below I wish to wax eloquent on how things look/are.

I will restrict this to one-dimensional objects like sticks moving along the direction of their lengths, directly toward or away from the observer. When a physicist talks about the length of something, here is what she means: measure the positions of the two ends of the object at *the same time*; the difference of those positions is the length. When you look at a stick, you are not observing the stick ends where they were at one time but you are seeing the farther end as it was sometime earlier than when you see the closer end. Of course, this does not matter if the stick is at rest, but if it is moving it does matter. For everyday moving sticks, there is no perceptible change in the apparent length of sticks because speeds are much less than the speed of light. But what if the speed is really big, let's say 80% the speed of light? Then, as I will shortly show, the effect is really big. But before we go into how long the stick *looks or appears*, we better be sure we understand how long the moving stick *is*. The result from special relativity, using the definition of length I gave above, is that the moving stick is shorter by a factor of $\sqrt{(1 - (v^2/c^2))}$, so if $v = 0.8c$ (i.e. 80% the speed of light), the length of the moving stick is only 60% its length when at rest. (This effect is called length contraction.) So now, figure 2.3 shows the situation if the stick is coming toward you (you are on the right). Light (red arrow) leaves the far end of the stick and does not catch up with the near end of the stick until the stick has gone a long way (four stick lengths) and now light from the far (red arrow) and near (green arrow) ends move forward to your eye. So the stick looks to be 5 times longer than it actually is and 3 times longer than if it were at rest! Now, if the stick is moving away from you, the situation is very different and is shown in figure 2.4. (The scales of the two figures are different; note the different rest lengths. I had to do this so the 'much-shorter' and 'much-longer' figures would be about the same size.) The moving stick is still 60% of its at-rest length, but now the near end moves away to 'meet' the light from the far end; the result is that the stick, as shown in figure 2.4, looks much shorter than it is. It now appears to be only $\frac{1}{3}$ the rest length or $\frac{5}{9}$ the actual (moving) length. Note that in neither case does the stick appear to be its actual length. So, maybe you can now understand why I often make a big deal about relativity being about *how things are, not how things appear*.

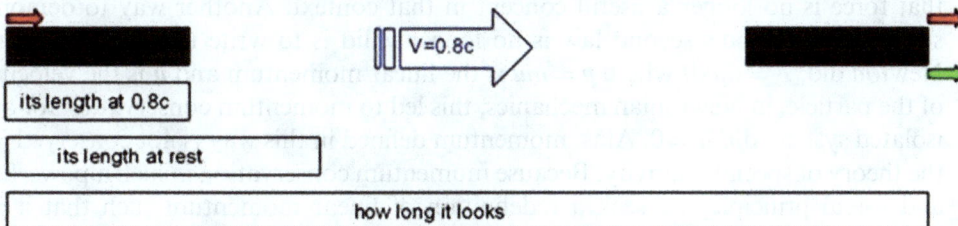

Figure 2.3. The appearance of a stick moving towards you at a high speed.

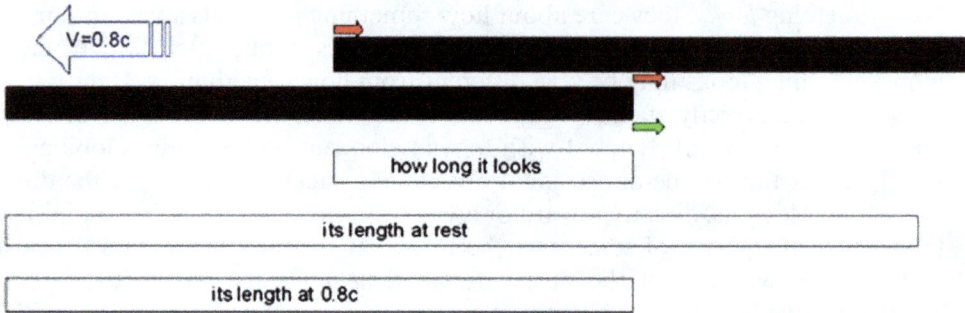

Figure 2.4. The appearance of a stick moving away from you at a high speed.

The above example was very simplified for sticks moving straight at or away from you. For more three-dimensional objects the analysis gets much more involved but leads to some fascinating results. A wonderful animation showing that a sphere moving by you at high speed *looks just like* a sphere can be seen at http://th.physik. uni-frankfurt.de/~scherer/qmd/mpegs/lampa_terrell_penrose_info.html. It looks like a sphere even though it is really very 'squashed' into a pancake shape.

2.7 Linear momentum, force, energy

The following question illuminates why the Newtonian definition of linear momentum is not useful because it is not a conserved quantity for an isolated system. Instead, linear momentum is redefined in such a way that it is, in special relativity, conserved and is approximately the Newtonian definition for small velocities.

Question: I understand the Lorentz transformation and how you can get time dilation and length contraction from it. What I cannot understand is why mass increases by a factor $\gamma = 1/\sqrt{[1 - v^2/c^2]}$.
Answer: Length and time are what we call kinematic quantities. Mass, force and, more importantly, linear momentum and energy are what we call dynamic properties. So, just like in an introductory physics course where, after we learn how to describe motion (kinematics), we next want to understand how motion can be changed; in classical physics this leads us to Newton's laws. What happens in relativistic physics is that we quickly find that Newton's second law, in the form $F = ma$ is no longer a true law of physics; that is, if two observers both measure the acceleration of a mass m they will get different answers for a, so that would mean that force is no longer a useful concept in that context. Another way to demonstrate that Newton's second law is no longer valid is to write the second law as Newton did, $F = \mathrm{d}p/\mathrm{d}t$ where $p = mu$ is the linear momentum and u is the velocity of the particle; in Newtonian mechanics, this led to momentum conservation for an isolated system, $\mathrm{d}p/\mathrm{d}t = 0$. Alas, momentum defined in this way is not conserved in the theory of special relativity. Because momentum conservation is such a powerful and useful principle, we seek a redefinition of linear momentum such that it is conserved for isolated systems and reduces to the old definition for small speeds.

If we define momentum as $p = \gamma m u$ we find that momentum is conserved in an isolated system and $p \approx m u$ for small u. So, you see, the gamma factor comes from redefinition of momentum, not redefinition of mass. Almost all introductory physics texts say that it is mass which increases, and this is certainly a possible interpretation of the new definition of momentum. I prefer to say that m is the inertial mass of an object at rest and that $p = \gamma m u$. However, there is nothing wrong with saying that the definition of momentum is Newtonian and the mass (inertia) turns out to depend on its speed.

Sometimes it is most convenient to say that m is the mass at speed v, is $m = \gamma m_0$ and I will sometimes do that; just be aware that in those instances the rest mass will be denoted m_0. With this definition of linear momentum, we can now rederive the work–energy theorem relativistically which means calculating the work done and equating it to the change in kinetic energy. The result is that $K = \gamma m c^2 - m c^2$ and, for small v, it is shown that $K \approx \frac{1}{2} m v^2$ (see appendix A). Now suppose that we rewrite the kinetic energy equation as $\gamma m c^2 = K + m c^2$. On the left we have some kind of energy and on the right we have kinetic energy plus some other kind of energy; if the particle is at rest, the 'some kind of energy' and the 'other kind of energy' are the same, $m c^2$. $m c^2$ is called the *rest mass energy* of the particle with mass m, energy which an object has by virtue of its having mass. $\gamma m c^2$ is the total energy m has, the sum of its kinetic and rest mass energies, $E = K + m c^2$. Finally, if you write $\gamma m c^2$ in terms of the momentum $p = \gamma m v$, it is easy to show that $E = \sqrt{(p^2 c^2 + m^2 c^4)}$.

Aha! We have just derived the famous equation $E = m c^2$! But, because of possible confusion with what m means, you have to be careful. With my meaning of m, the rest mass of the particle, $E = m c^2$ means the energy of a particle at rest. If by m you mean the increased mass with velocity, then $m = \gamma m_0$ where m_0 is the rest mass, and $E = m c^2$ means the total energy of a particle with mass m. This confusion is the source of a great many questions I have answered wondering why a photon which has no mass can have energy.

Question: Einstein's famous $E = m c^2$ doesn't seem to hold for a photon which is massless but has energy. What am I missing?

Answer: I often get this question. It originates with taking a famous equation and not understanding when it is applicable. $E = m c^2$ is the energy of a particle of mass m at rest; a photon is never at rest and therefore this equation is not applicable to it. The energy of any particle is $E = \sqrt{[m^2 c^4 + p^2 c^2]}$ where p is the linear momentum. Note that if $p = 0$, the particle is at rest and indeed $E = m c^2$. If $m = 0$ then $E = pc$. Massless particles have momentum. The only massless particle we know is the photon which has an energy $E = hf$ where h is Planck's constant and f is the frequency of the corresponding electromagnetic wave. So the momentum of a photon is hf/c.

Finally, Newton's second law takes the same form as before except momentum is redefined. It is interesting to now do the simplest Newton's second law problem, a mass m experiencing a constant force F. Recall that the nonrelativistic solution to this problem is $x = \frac{1}{2}(F/m)t^2$ and $v = (F/m)t$.

Question: If you could drop a rock down an infinitely deep well with a constant gravitational 'pull', what formula would describe its velocity in terms of time falling? I know it starts as $v = at$ and approaches $v = c$, but what does it do in between?

Answer: Let us stay away from gravity since the definition of a uniform gravitational field is problematical. But, I think what you are interested in is what is the velocity if the applied force on a mass m is constant. I know that an object with zero net force has $dp/dt = 0$ where $p = mv/\sqrt{[1 - (v/c)^2]}$. Now, I am going to define a constant force F to be one for which the rate of change of momentum is constant, that is $dp/dt = F$ where F is the constant. As noted by the questioner, $v \approx Ft/m$ for small time and $v \approx c$ for large time. It is very easy to integrate $dp/dt = F$ to get momentum as a function of time, $p = Ft$. Putting in what p is in terms of v and solving for v, I find $v/c = (Ft/(mc))/\sqrt{[1 + (Ft/(mc))^2]}$. This function has the correct properties at small and large t and is shown in the graph in figure 2.5. This is the correct $v(t)$ for a constant rate of change of momentum F. **Note added later:** Someone expressed interest in the position as a function of time for this problem. This is straightforward to do by integrating $dx = vdt$. Doing this I find $x = (mc^2/F)(\sqrt{[1 + (Ft/(mc))^2]} - 1)$. Note that this has the expected properties that for small time, $x \approx \frac{1}{2}(F/m)t^2$, and for large time, $x \approx ct$. (I assumed $x = 0$ at $t = 0$.)

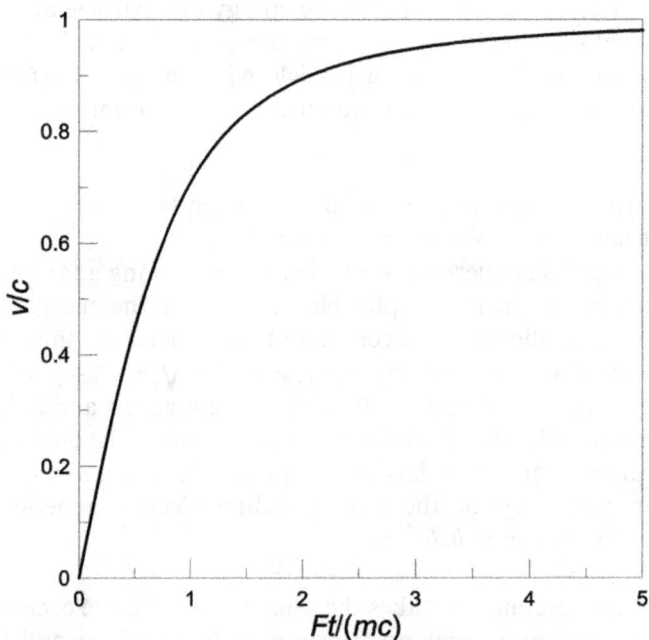

Figure 2.5. A graph showing time dependence of velocity for a constant force applied.

A related problem, which I will do in section 2.10, is the motion of a particle experiencing constant weight mg which differs from constant force in that m varies with v.

2.8 The universal speed limit

We have already seen an indication that the speed of light is the fastest possible speed in the Universe. In section 2.2 when the velocity addition theorem was discussed, we found the correct expression was $v' = (u + v)/[1 + (uv/c^2)]$. This equation says that you cannot find something going faster than c provided that both u and v are less than c; for example, suppose that you are going with the speed $0.9c$ and somebody is coming toward you with the speed of $0.9c$. Then the speed you see him approaching you is $v' = (0.9 + 0.9)c/(1 + 0.81) = 0.994c$. So, you cannot get above light speed by velocity transformation. But, why not just brute force it, pushing really hard for a really long time. Chuck Yeager broke the sound barrier so why can't we break the light barrier?

Question: I know that nothing can travel at or faster than the speed of light. But, just simply why? What equations or whatever says no?

Answer: Because the mass of an object, that is its inertia, increases as the velocity increases. Therefore it gets harder and harder to accelerate it as it goes faster and faster. The expression for the mass of an object m as a function of its velocity v is $m = m_0/\sqrt{(1 - (v^2/c^2))}$ where c is the speed of light and m_0 is the mass when it is at rest. Note that as v approaches c, m approaches ∞ so it is impossible to push beyond c. Another way to look at it is from the perspective of energy. The energy of a particle is $E = mc^2 = m_0c^2/\sqrt{(1 - (v^2/c^2))}$, so the energy required to accelerate the mass to the speed of light is infinite and there is not an infinite amount of energy in the Universe.

So, you can forget about brute-forcing it against the 'light barrier'. But, folks really want to go faster than c and keep coming up with clever scenarios (which don't work!).

Question: Let's say I have a metal rod about a half an inch thick and 300 000 km long. Then say I give one end of said rod a mighty whack with a hammer, propelling it forward by one inch in a mere fraction of a second. My questions is, wouldn't the impact of my hammer cause the other end of the rod to move forward one inch just as rapidly as the end where I whacked it? And would this violate Einstein's law that states that nothing can move faster than c? Or would the far end of the rod have to wait 1 s after my whacking my end before moving forward by one inch?

Answer: Have you thought about the implications of your question? I figure the mass of the rod would be about 10^{10} kg. Suppose that you exert a constant force such that after 0.1 s it is moving with a speed of about 0.5 m s^{-1}; it would have moved about an inch in this time. The force is the change in momentum divided by the elapsed time so, roughly speaking, the required force is about 10^{12} N. Where are you going to get such a force? Anyhow, to the meat of your question: no, the other end would not start moving instantaneously. It could not begin moving until at least 1 s later than your end started moving for the reason you state: no information can travel faster than c. In reality, it would be much longer than 1 s because your 'mighty whack' will compress the rod and this compression will move with the speed of sound in the metal and this compression is what travels to the other end to move it.

2.9 Energy from mass, mass from energy

Where do we get our energy? Can we ever hope to see mass turned into energy? Or do we see it happening all around us every day?

Question: I know Einstein said $E = mc^2$ and basically all matter can be equated to some quantity of energy; then why do we go to the gas station to fill our cars? Why can't we use garbage, which is mass and has energy, to power our cars? How can we convert matter to energy? I know we can burn gasoline to use perhaps $\frac{1}{4}$ the heat content in the form of expanding gas to apply pressure to the piston in the engine. Has anyone invented a converter that changes matter to energy yet? We eat food and basically run on sugar which fuels a chemical based process. Any other matter converters?

Answer: Most of the energy mankind uses comes from chemistry. Burn coal or gasoline, for example. When you eat and metabolize food, chemistry is going on. The energy which is extracted comes from—guess what—mass! For example, when you burn coal the main thing which is happening is that carbon is combining with oxygen to form carbon dioxide. One carbon dioxide molecule has a smaller mass than one carbon atom plus one oxygen molecule. So, chemistry is the best known example of your 'matter converter'. The problem is that an extremely tiny fraction of the mass is converted to energy, something like 0.00000001%, so chemistry is a very inefficient source of energy. Now, to get more efficiency we have to work not with atoms but with nuclei of atoms. If a heavy atomic nucleus can be induced to split (fission), the mass of the fragments is smaller than the mass of the initial nucleus by an amount much bigger than with chemistry, something like 0.1% which is a huge improvement over chemistry; this is how nuclear reactors work. Also from nuclear physics, you can take very light nuclei and make them combine (fusion) and get something like 1% of the mass converted into energy; this is how stars work and so, you see, solar energy comes from 'matter converters' too and so does wind energy since the Sun is the energy which causes winds to blow. If you want to get 100% efficient you have to go to particle–antiparticle interactions in particle physics. When an electron and its antiparticle the positron meet, their mass completely disappears and all the energy comes out as photons. Did you ever see the *Back to the Future* movies? Doc came back from the future where they had invented a small appliance called 'Mr Fusion' to do what you want, to convert garbage into the huge amount of energy needed to power the time machine.

And, the other way around, matter from energy.

Question: Conversion of mass to energy (fission) has been demonstrated many times in laboratory and field tests. Has conversion of energy to mass also been demonstrated in laboratories?

Answer: Yes. A couple of examples:
  • A very energetic photon (massless) can spontaneously turn into an electron–positron pair; this is called pair production.

• The mass of a nucleus is always less than the sum of all the constituent proton and neutron masses. Suppose you remove a neutron from a nucleus; it will take work because that neutron is bound in the nucleus. Hence, the final system of neutron and the original nucleus minus one neutron has a greater mass if both objects are at rest. So let's just say that the nucleus, having a mass smaller than the sum of its parts, is an example of converting energy into mass because there is more mass after you dis-assemble it by adding energy (doing work).

2.10 These are a few of my favorite things

As in chapter 1, I collect here a few of my favorite miscellaneous answers about special relativity.

Question: According to special relativity, as an object accelerates, gets closer to the speed of light its mass increases. Since the mass increases the gravitational force increases, so it is not a constant force like you have worked out before, but an increasing weight problem.

Answer: To know how the mass varies (if you interpret relativistic momentum that way), just calculate $m_0/\sqrt{(1-\beta^2)}$ where m_0 is the rest mass and $\beta = v/c$. It occurs to me, though, that it might be of interest to redo that calculation for a force which is not constant but which has a value $mg = m_0 g/\sqrt{(1-\beta^2)}$ where g is the acceleration due to gravity in whatever strength field you wish to examine. I will not give all the details here, just the results. We start by integrating the relativistically correct form of Newton's second law:

$$m_0 g/\sqrt{\left(1-\beta^2\right)} = dp/dt = (d/dt)\left[m_0 v/\sqrt{\left(1-\beta^2\right)}\right]$$

integrated gives

$$gt/c = \frac{1}{2}\ln[(1+\beta)/(1-\beta)]$$

solved for β gives

$$\beta = (1 - \exp(-2gt/c))/(1 + \exp(-2gt/c));$$

put this into the mass and get

$$m/m_0 = 1/\sqrt{\left(1-\beta^2\right)}.$$

These are plotted in figure 2.6. Note that the result for a constant force $m_0 g$ is shown as a dashed line for comparison. Potential energy is not a useful concept here. Note that the time to get to near c, $t \approx 3c/g$, is about 2.9 years for $g = 9.8$ m s^{-2}. At that time the moving mass is about 10 times greater as shown in figure 2.7. A more lengthy discussion of a mass in a uniform gravitational field (including general relativity) can be seen here.

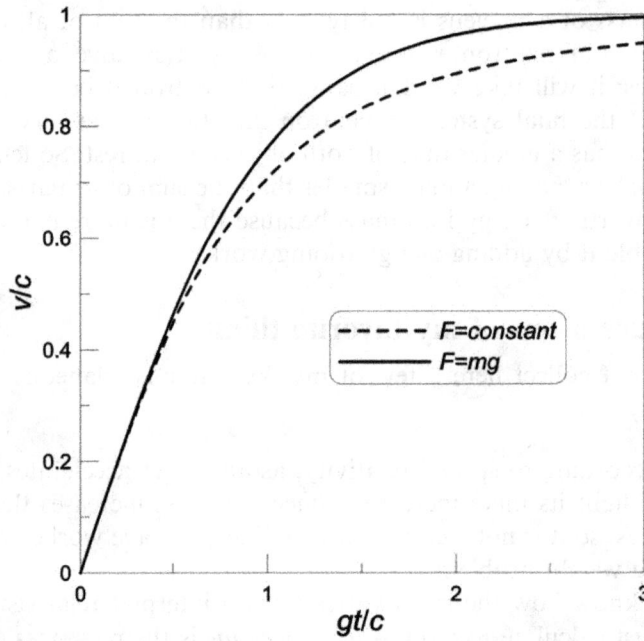

Figure 2.6. A graph comparing time dependence of velocities for a constant force (m_0g) and for a gravitational force (mg).

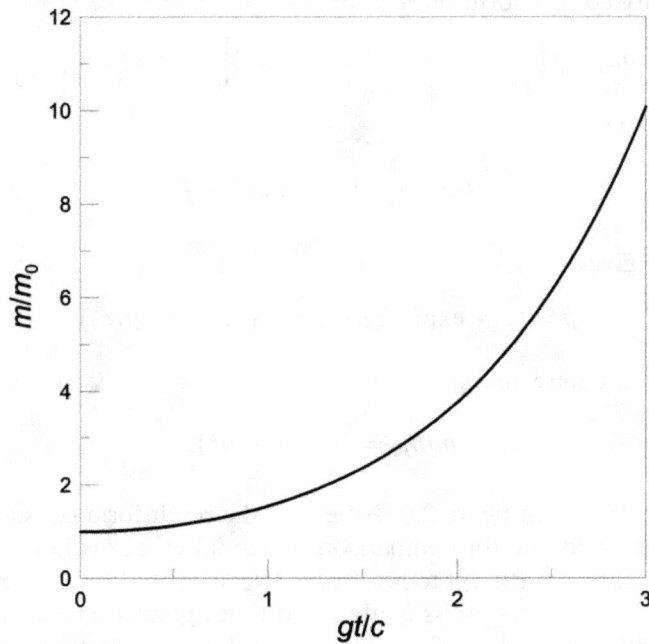

Figure 2.7. A graph showing the time dependence of the mass falling in a uniform gravitational field.

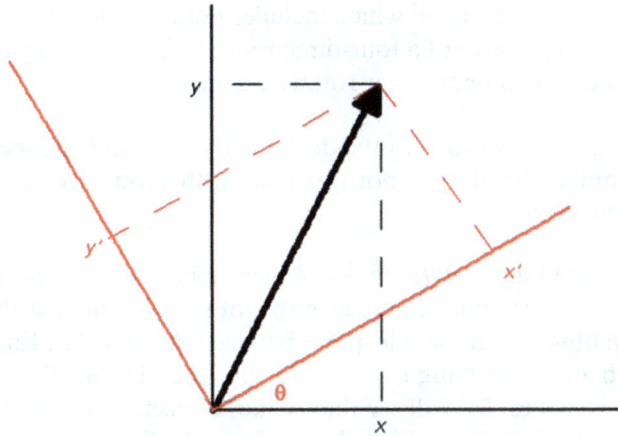

Figure 2.8. Rotated Cartesian coordinate systems.

The notion that relativity implies that time is like a fourth dimension and the form of the Lorentz transformation, mostly neglected in this book, are touched on in the next question.

Question: I read *A Brief History of Time*, but it has left me confused. Can you please explain the 4 dimension rule. How time and distance are related.

Answer: You need to learn the theory of special relativity. It requires only algebra to understand its basic concepts. Essentially, if we demand that the speed of light in vacuum is the same for all observers (by now a well-documented experimental fact), the inescapable conclusion is that space and time are not separate entities but are 'entangled'. The 'entanglement' is much like that of the three spatial dimensions. An example is the rotation of a coordinate system shown in figure 2.8. Note that the components of the vector in the black coordinate system (x,y) are not the same as in the rotated (red) coordinate system (x',y'). In fact, (x',y') depend on what (x,y) are:

$$x' = x \cos \theta + y \cos \theta$$
$$y' = -x \sin \theta + y \sin \theta.$$

The rotation 'mixes up' the spatial dimensions. It turns out that in special relativity, the transformation between space and time if one system is moving with speed v with respect to the other looks very much like a rotation in 'four-dimensional space' because x (the direction of the relative velocity) and t get mixed up:

$$x' = \gamma (x - vt)$$
$$t' = \gamma \left(t - vx/c^2 \right)$$
$$\gamma = \left(1 - (v/c)^2 \right)^{-\frac{1}{2}}.$$

The fact that this transformation mixes up these two dimensions leads us to recognize that space and time are no longer separate unrelated concepts. Instead, we should think of 'space–time' which includes both time and the three dimensions of space, thereby suggestive of a four-dimensional world. Moving from one inertial frame to another is analogous to a rotation is space–time.

One of the cornerstones usually introduced early in a study of special relativity is the notion of simultaneity. I have not touched on this topic, so the question below addresses this omission.

Question: I'm reading *Relativity* by Albert Einstein and in it he says that simultaneity is relative and uses the example of two lightning strikes at points 'A' and 'B' and an observer at the mid-point between the two 'M'. Einstein states that an observer on a train moving from 'A' to 'B' observing the lightning strikes just as the observer reaches 'M' will see that strike 'B' occurs before 'A' (as the train is moving toward the light from 'B' and away from the light coming from 'A'). Does this mean that an event only becomes real when the light from it reaches the observer? Or could one legitimately say 'These two events happened simultaneously, but I saw the flash from 'B' before I saw the flash from 'A'.'?

Answer: The relativity of simultaneity is not how it *looks*, but rather how it *is*. In relativity you have to be careful how you define a time interval so that you correct for such things as the time it takes light to reach you. The converse is also true, a distant star is seen right now but as it was long ago. I always find the following example more convincing than the lightning at the ends of the train example. In the center of one of the cars there is a flashbulb which flashes (see figure 2.9). Light reaches the front and rear walls of the train at the same times. However, an observer on the side of the track watching the car go by sees the rear wall come forward to meet the oncoming light and the front wall running away from the light coming at it. Hence the event corresponding to the flash at the back wall occurs earlier for the observer at rest. (It is crucial to realize that both observers see all light moving with the same speed.)

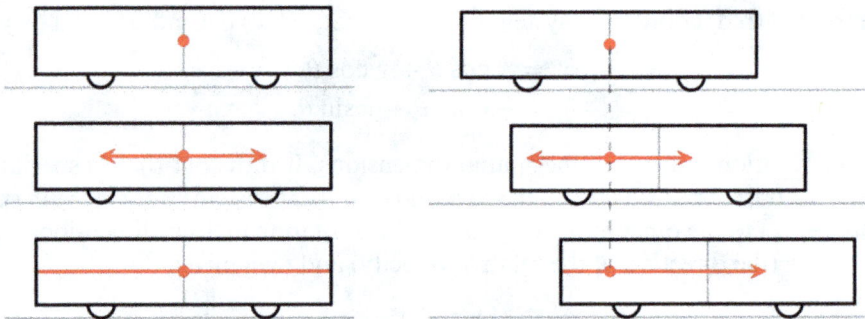

Figure 2.9. Left: observer on the train. Right: observer by the tracks.

Finally, when the Large Hadron Collider came on line at CERN, there were a number of panicy outcries fearing that something catastrophic would happen, like a mini black hole being created and sucking in the whole Earth. Here are a couple of questions along those lines.

Question: I understand that matter gains in mass as you accelerate it towards the speed of light, and that as you approach the speed of light, mass tends toward infinity, meaning you would need an infinite force to 'push' it beyond the speed of light, and that's one of the reasons why nothing can exceed the speed of light. However, I also understand that with increased mass comes increased gravity, so... When they accelerate those sub-atomic particles to velocities approaching the speed of light at CERN, why don't they suddenly start exerting lots of gravity and sucking the entire CERN campus toward them?

Answer: First, a little perspective. If I were to bring a baseball into the CERN campus, I think you would agree that it would not 'suck in' everything grav- itationally. So, an accelerated proton better be a whole lot more massive than a baseball! Suppose the speed of the protons were 99.99999999% (that's ten nines) of the speed of light. Then you can calculate the mass of a proton going this fast as $m = 1.7 \times 10^{-27}/\sqrt{(1 - 0.9999999999^2)} = 1.2 \times 10^{-22}$ kg. The proton has increased mass by a factor of about 70 000 and that is still far from a baseball mass.

Question: This lady lost her legal battle to have the Large Hadron Collider shut down. She believed that the atom smasher could create a black hole and suck up the Earth. The courts sided with the scientists. The scientists said it's not possible, and even if it did create a black hole, it would be a micro black hole and collapse in on itself. That seems to go against the laws of physics and quantum physics. We now know that matter sucked up by a black hole is permanently imprinted on the black hole's surface and the size of the black hole increases. Did I miss something?

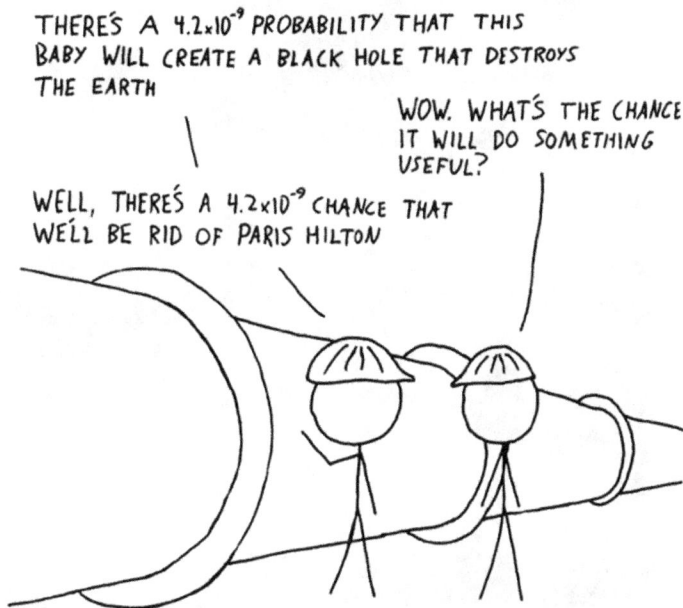

THERE'S A 4.2×10^{-9} PROBABILITY THAT THIS BABY WILL CREATE A BLACK HOLE THAT DESTROYS THE EARTH

WOW. WHAT'S THE CHANCE IT WILL DO SOMETHING USEFUL?

WELL, THERE'S A 4.2×10^{-9} CHANCE THAT WE'LL BE RID OF PARIS HILTON

Answer: I do not normally answer astrophysics questions, but I can deal with this one, I think. First, some cosmic rays (radiation which strikes Earth from space) have energies much greater than the proton energies in the LHC and if a black hole could be created and have sucked in the whole Earth, that would have happened long before we evolved. Second, you did miss something—Hawking radiation whereby a black hole can radiate energy. For something to be 'sucked into' the black hole, it must come within the Schwartzchild radius which is

$$2Gm/c^2 = 6.67 \times 10^{-11}m/(3 \times 10^8)^2 \approx 1.5 \times 10^{-27}m$$

where m is the mass of the black hole. The maximum kinetic energy of each proton in the collider is 7 TeV which is about equal to the mass energy of 7000 protons, so the heaviest black hole they could make would have a mass of about 14 000 protons, about 2.3×10^{-23} kg; the corresponding Schwartzchild radius is 3.5×10^{-50} m! It seems to me that even if this black hole never evaporated, it could go a really long time before it got close enough to anything to suck it in.

From Newton to Einstein
Ask the physicist about mechanics and relativity
F Todd Baker

Chapter 3

General relativity

3.1 Overview

There is a story about Albert Einstein, probably apocryphal, but amusing and instructive. As you probably know, his first job was as a clerk in the Swiss Patent Office and he could easily complete all his responsibilities quickly and then spend the rest of the day thinking about physics. One day, while watching a workman paint a building across the square, the workman fell from his ladder. Einstein thought to himself, 'there is no experiment which he can do which could distinguish whether he was in free fall in a gravitational field or was in an inertial frame in empty space'. This is, in essence, the *equivalence principle* which, along with a *generalized principle of relativity*, provide the basis for the general theory of relativity. General relativity is an extremely mathematical theory, well beyond the scope of this book to discuss. But, as for much of physics, a good qualitative understanding based on comprehensible principles can be achieved. To that end, let us carefully state the two principles I would consider essential to understand.

- The principle of relativity is now stated as: the laws of physics are the same in all frames of reference, not just inertial frames.
- The equivalence principle states that there is no experiment you can do which can distinguish whether you are in an accelerating frame of reference, acceleration a, or in a gravitational field whose local gravitational acceleration is a.

General relativity is the theory of gravity. It tells us why masses attract each other with what appears to be a force but is actually a deformation of the four-dimensional space–time. Euclidean geometry is no longer valid, the shortest distance between two points is no longer a straight line.

3.2 Examples

The equivalence principle can be used on everyday problems, not just esoteric gravitational scenarios, as illustrated in the following example. (This example also demonstrates that sometimes *The Physicist* can get it wrong!)

doi:10.1088/978-1-6270-5497-3ch3   3-1

Question: Suppose we have Bob the astronaut sitting without a space suit in a spaceship full of air on a mission to Mars. Bob is very fond of balloons and is holding on to a nice, big, red helium balloon via a piece of string. Bob is sitting facing the front of the spaceship. Mission Control decides to slowly accelerate the spaceship. In which direction will the balloon move relative to Bob? Why?

Answer: I do not like this question because the reason a helium balloon floats is that the buoyant force, which floats it, arises because the pressure in the air is bigger underneath the balloon than above it; on a mission in empty space, the pressure everywhere in the cabin is the same and so the balloon would not go up! Let's just have Bob ride in an accelerating car right here on Earth. There are three forces on the balloon, its weight W, the buoyant force B, and the tension in the string T. B and T are both vertical, and so, for the balloon to have an acceleration in the direction of the acceleration a of the car, T must have a horizontal component in the direction of a. All this is shown in figure 3.1. Therefore the balloon will move backward opposite the direction of the acceleration. In the spaceship where there would be no buoyant force and no weight, the balloon would appear to accelerate backwards until the string was straight and 'horizontal'.

'Enhanced' answer: It has been pointed out to me (thanks to Michael Weissman at *Ask the Van*) that my answer would be correct only if there were no air in the spacecraft (contrary to the stipulations of the question) and the explanation of the balloon in the accelerating car was flat-out wrong! What I failed to think about was that if there is air in the cabin, forward acceleration will cause the pressure at the rear of the cabin to be greater than at the front; therefore there would be a buoyant force on the balloon from back to front. So, if the mass of the balloon is less than the mass of an equal volume of air, the direction the balloon would move would be forward, not rearward. If you are familiar with the equivalence principle, it is even easier to understand. The equivalence principle states that there is no experiment you can perform which can distinguish between an accelerated frame and being in an unaccelerated frame in a uniform gravitational field with the same acceleration (due to gravity). So, if the acceleration of the ship were g, the

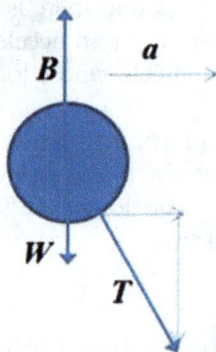

Figure 3.1. Forces on a buoyant balloon.

balloon would have to behave just the same as it would on Earth except forward now would play the role of up on Earth, so the balloon would go forward just like the balloon on Earth would go up.

For the car with acceleration a on Earth, there would be a forward buoyant force whose magnitude would be $\rho_{air}Va$ in addition to the forces shown in figure 3.1 and $B = \rho_{air}Vg$, $W = \rho_{helium}Vg$; here V is the volume of the balloon, ρ_{helium} and ρ_{air} are the densities of the air in the car and the helium in the balloon, respectively. Without going into detail, I find

$$\begin{aligned} \tan\theta &= \left(\rho_{air}Va - \rho_{helium}Va\right)/\left(\rho_{helium}Vg - \rho_{air}Vg\right) \\ &= (a/g)(\rho_{air} - \rho_{helium})/(\rho_{helium} - \rho_{air}) \\ &= -a/g. \end{aligned}$$

Here θ is the angle T makes with the vertical in figure 3.1; the fact that $\tan\theta$ (and therefore θ) is negative means that the balloon will move forward, not backward as in the figure.

One of the first experimental verifications of general relativity came from observing light being bent by a strong gravitational field, one of the key predictions of the theory. In 1919 observations of starlight being bent by the gravity of the Sun during a solar eclipse were made and shown to be consistent with general relativity predictions. Using the equivalence principle, it is fairly easy to understand why gravity bends light even though it does not have any mass, as illustrated by the following question.

Question: My 11 year old has asked me, 'Does gravity bend light?' I did high school physics about 20 years ago, so am very rusty and not up to date with current thought. I have looked at this discussion topic on the archives of many forums, but have ended up very confused by the differing opinions/explanations. I would really like an answer which is easy to explain to a child, but yet not so simplistic that it is inaccurate. Am I asking the impossible? She has read about the nature of light and also about gravity, and can't understand how light can be affected by gravity, when it has no mass. Is it because photons are energy and so can be used instead of mass? Or is it that the gravitational pull around massive stars affects the 'space' around it and light just follows the stretched paths? If this is true, how can some authors say that the light is still moving in a straight line even when it is following a curved path?

Answer: Here is one explanation, probably the easiest for your daughter to understand: light being affected by gravity is a result of the equivalence principle in general relativity. This states that there is no experiment which you can perform to distinguish between your being in a gravitational field or in an accelerated frame of reference. Thus, for example, imagine that you are in an elevator which accelerates upward; if light enters through a hole in the side of the elevator it will clearly appear to fall like a projectile because of the acceleration of the elevator. So, the same thing will appear to happen in a gravitational field the acceleration due to which is exactly the same as the acceleration of the elevator. Hence, light will 'fall' in the Earth's gravitational field with an acceleration of $9.8 \, \mathrm{m \, s^{-2}}$.

Here is another: If we look at the world as having a Euclidean 'flat' geometry and watch a ray of light pass a very massive object, we see the light bend. But, the way that general relativity describes the world says that, if we are in the vicinity of a massive object, the space itself is not Euclidean but is curved; in this space the light follows a 'straight line' in that non-Euclidean geometry. Think about moving on a two-dimensional space like the surface of the Earth. The shortest distance between two points on Earth is not really a straight line, but it is a straight line in that space.

So, if gravity bends light, doesn't that mean that it exerts a force on it and so you can speed it up if it is traveling toward a mass? No, it doesn't as the following question shows.

Question: Doesn't the fact that a black hole can bend light prove that something can travel faster than the speed of light? As light is pulled toward the black hole it would accelerate, since it is already traveling at the speed of light the moment it started moving toward the black hole it would be going faster than the speed of light would it not? Just curious.

Answer: No, the light does not speed up as it falls into the black hole. What happens is that, as you would expect, it gains kinetic energy as it falls but light's energy is all kinetic. But, your idea of kinetic energy is probably $\frac{1}{2}mv^2$, but this obviously cannot be true for light since it has no mass. The energy of a photon is hf where h is Planck's constant and f is the corresponding frequency of the electromagnetic wave. So, what happens when a photon gains energy is that the frequency increases; this is known as a gravitational blue shift (the color of the light moves to shorter wavelengths) and happens when a photon approaches any massive body, not just a black hole. A photon moving away from a large mass experiences a gravitational red shift, a loss of energy.

Another consequence of the equivalence principle is a gravitational time dilation, this is in addition to the time dilation in special relativity. I will not go into the details here, but it is completely equivalent to the gravitational red/blue shift. Essentially, clocks run slower as a gravitational field increases, so a clock upstairs will run faster than one in the basement. Both gravitational and velocity time dilations must be taken into account if extremely accurate time measurements are required for a particular technology. GPS systems are one such technology; the time of transit of radio signals between satellites and you are the basis of locating you and corrections for time dilation (both kinds) are imperative for the amazing accuracy of GPS. Here is a question about gravitational time dilation which will give you a feeling for the magnitude of the effect.

Question: How much faster does time pass, out in the middle of nowhere in a space that is not effected by gravity? Like how much faster will time go compared to a clock on Earth if we stuck another clock out in a magical spot in space where the gravitational pull of galaxies would not effect it?

Answer: The equation for gravitational time dilation is $\sqrt{(1 - (GM/(Rc^2)))}$ where G is the universal gravitational constant, M the mass of the object, R the distance from the object and c the speed of light. At the surface of the Earth this is approximately $1 - 7 \times 10^{-10}$ and for $R = \infty$ it is 1. So time passes about 7×10^{-8} % faster in empty space.

Science popularizations often use some variation of the following model to explain the bending by mass of space–time. Imagine a bowling ball on a trampoline; the trampoline is pushed down leaving it not flat. Now imagine putting a marble on the edge of the trampoline; it will move toward the bowling ball, not because of some force but because the trampoline is warped. I think it is a really good simplistic way of qualitatively understanding what is going on. But many questioners at AskThePhysicist.com are thinkers and bothered by this model. How can you employ gravity in an example which purports to explain gravity, they ask. How can you possibly describe warping of four-dimensional space by a two-dimensional model? The next question demonstrates my take on this.

Question: My question is about gravity. In the depictions I have seen of the Einstein model of gravity, planets and stars are shown as depressing a plane of space–time into a well like depression into which other objects tend to fall. I am ok with this depiction. However, it seems to make the assumption that space is a plane and has only two dimensions. When I observe the Universe, I see three dimensions. It would seem to me that these 'gravity wells' should exist in three dimensions not the two generally depicted. In the two-dimensional illustrations, these gravity wells seem logical and simple—the Sun for example presses down to form a depression in the two-dimensional plane and the Earth falls in towards it in an orbit. But space is not two-dimensional. These 'wells' or depressions should exist in an infinite number of orientations in a three-dimensional space. Why are they only shown as if the fabric of space is like a sheet of paper, in two dimensions and not in an infinite number of orientations as would be the case in a three-dimensional space?

Answer: My stock answer to this kind of question is that the 'trampoline illustration' of deformed space–time is meant to be a cartoon to illustrate the idea, not an accurate rigorous representation of the theory of general relativity. You must not take it too seriously or literally. It is also practically impossible to draw a picture of deformed three-dimensional space. To draw deformed two-dimensional space is easy because you use the third dimension to show the deformation. To draw a deformed three-dimensional space would require a fourth spatial dimension which cannot be drawn. To make things even worse, what is really deformed in general relativity is space–time, so you would have to somehow envision drawing a four-dimensional figure in five dimensions.

An important issue in modern astrophysics and cosmology is the quest for a theory of quantum gravity.

Question: My question is in regards to relativity and quantum mechanics. I've heard physicists say on TV programs and in books that relativity and quantum mechanics are incompatible and describe two different phenomena. Do these two theories contradict one another? Is the Universe contradictory or is there anything in the Universe that can definitively be defined or described as a contradiction?

Answer: You have to be careful that when somebody refers to 'relativity', you know what they are referring to. The theory of special relativity, which predicts things like time dilation, length contraction, $E = mc^2$, etc, is perfectly compatible with quantum mechanics; there is something called relativistic quantum mechanics. What is probably being referred to by your sources is the theory of general relativity which is the best current theory of gravity. No one has been successful in developing a theory of quantum gravity. I would not so much describe them as 'incompatible' as I would call them non-unified'.

The following is one of the most interesting questions I have answered. This is for real connoisseurs of the equivalence principle!

Question: Suppose you have radiation detectors fixed on the ground on Earth. Will they detect radiation coming from a charged particle in free fall near them? The first answer that comes to mind is: Yes, they will detect radiation because the particle is accelerated, and electrodynamics predicts that accelerated charges must radiate in this situation. According to the Equivalence Principle, this situation is equivalent to detectors fixed on an accelerated rocket with acceleration g moving in the outer space and far away from the influence of other bodies. If the answer to the previous question is yes, then the detectors on the rocket should also detect radiation coming from a charge in free fall as observed by the reference frame of the rocket. But a charge in free fall in this reference frame is at rest in the inertial reference frame fixed with respect to the distant stars, and a charge at rest in an inertial frame should not radiate. Is it possible that detectors fixed on the rocket detect radiation but detectors at rest in the inertial frame do not? Is radiation something not absolute, but relative to the reference frame?

Answer: This is a fascinating question and points to an experiment which would seemingly violate the equivalence principle. The answer to your first question is an unequivocal yes, an electric charge accelerating in free fall in a gravitational field radiates electromagnetic waves, an electric charge not accelerating does not radiate. But, suppose that you are falling along with the charge; relative to you the charge is not accelerating and therefore not radiating. Or, equivalently, suppose that you are in a spaceship in empty space with your rockets turned on. If you release an electric charge inside, it will 'accelerate' toward the rear of the ship and therefore radiate because the equivalence principle states that there is no experiment you can perform which can distinguish between the accelerating frame and a static gravitational field. However, the charge will move with constant speed relative to an inertial observer nearby and therefore not radiate. In both cases we have an electric charge both radiating and not radiating, a seeming paradox. Although I had not heard of this paradox before, apparently it has been a topic of many articles. The most recent of these, by Almeida and Saa, has evidently laid the paradox to rest. They demonstrate in this article that observers for whom the charge is not accelerating '...will not detect any radiation because the radiation field is confined to a space–time region beyond a horizon that they cannot access...' and '...the electromagnetic field generated by a uniformly accelerated charge is observed by a comoving observer as a purely electrostatic field'. Like all 'paradoxes' in relativity, there is not really a paradox; rather a radiation field in one frame may be a static field in another. Basically, you nailed it when you said 'radiation [is] something not absolute, but relative to the reference frame'.

One of the predictions of general relativity is that disturbances like supernovae or black holes rotating around massive stars result in gravitational waves. So it is natural to ask what the speed of gravity itself is. For example, if all of the Sun's mass suddenly disappeared, how long would it be before we knew it, before our trajectory became a straight line instead of our orbit? Interestingly, this speed has never been measured. The following question addresses this point.

Question: What is the speed of gravity? If you don't have an exact speed calculated, is it faster than the speed of light?

Answer: No one has ever measured the speed at which the gravitational force propagates. In Newtonian physics we calculate orbits by assuming that the forces are instantaneously transmitted, but nobody believes this—it just works pretty well because realistic propagation times are bound to be very short over the distances of interest. A better theory of gravity is general relativity. Here the theory predicts that gravity is due to distortions of space–time caused by the presence of mass and these distortions are predicted to propagate with the speed of light. In general, all the physics we know forbids any information propagating faster than light speed. Finally, we believe that a theory of quantum gravity will someday be found in which the quanta which transmit the force will be massless particles called gravitons (similar to photons which transmit the electromagnetic force). Massless gravitons, like all massless particles, would have a speed equal to the speed of light so the force which they transmit would move with that speed. There is a very enlightening essay on John Baez's blog.

Finally, I need to expose the rebel in me! Cosmology, for which general relativity is gospel, is focused on two additional important questions, puzzles if you like, *dark matter* and *dark energy*. As the answer to the final question of this chapter shows, my take on these questions is decidedly not mainstream. So, here I am, like many of my questioners, an outsider who looks critically at something at which he is not expert.

Question: We've all been reading about dark matter and dark energy for some years now and I believe you've said that you (among many others) are not yet persuaded that dark energy and dark matter exist. If matter and energy, as traditionally and conventionally understood, comprise only a very small part of the substance of the Universe, does it follow that classical mechanics, thermodynamics, relativity and quantum theory, etc, correspondingly apply only to that (seemingly) tiny aspect of the world? Is there any reason to think that the laws and theories of physics that humanity has discerned to date would apply also to dark matter and dark energy?

Answer: First, a disclaimer: as I state on the site, I am not an expert in astrophysics, astronomy, or cosmology, so you can take my opinion with at least a grain of salt or ignore it altogether! I would not say 'many others'! Most astrophysicists, astronomers and cosmologists talk about dark matter as if it is surely there but just not directly observed yet. My own point of view is that I need to see some direct evidence before I accept that such a thing really is there; there is lots of indirect evidence of dark matter—the dynamics of galaxies, the time when galaxies first began to form, to name a couple—but it is altogether possible that we do not understand gravity as well as we assume that we do. The best theory of gravity, general relativity, makes many assumptions which are not necessarily true over really large distances. If this were the case, maybe dark matter is the 21st century equivalent of the lumeniferous æther and is something we are looking for in vain because there is no such thing. There are lots of good ideas about what dark matter might be (including WIMPS, for which some evidence has recently been observed in the observed excess of high-energy positrons) and I will be happy to accept experimental evidence when it happens. Dark energy is a different matter in that it has not caused a search for some 'stuff'. There is already a place for dark energy in general relativity, known as the cosmological constant; in other words, many cosmologists do have the point of view that dark energy does result from an incomplete theory of gravity.

Chapter 4

Wacky questions: sci-fi, super heroes, computer games, fantastic weapons, etc

4.1 Wacky questions

I love these kind of questions. And I love it that gamers, trekkies, sci-fi movie fans, and the rest stop to think about whether the things they encounter could ever really be possible. Mostly, it is just a matter of looking at the realities of either classical mechanics or relativity to see how crazy some of the assumptions are! For the most part, these questions stand on their own, do not require any additional commentary.

Question: I'm hoping you can settle a bet between my father and myself. We are both movie buffs, and work together. While working we discussed the Ah-nuld movie *Eraser*. I'm going to make the assumption you have never seen the movie which revolves heavily around the manufacture of man-portable super-railguns that fire aluminum rounds at 'close to the speed of light'. Being firearms enthusiasts as well, we discussed the flaws in their idea (beyond the fact no technology in the near future would allow such a weapon as the one depicted to be made, let alone one man-portable). Where we came to a disagreement was why for other reasons it would be useless. While we agree that the mooks Ah-nuld tears through as he does in every movie would be out of luck, in practical situations the weapon would be useless but we disagree why. My father believes that a 4 gram bullet (our assumptions) traveling 80% of c would immediately flatten out and quickly lose all of its kinetic energy due to its low mass and aluminum's relatively soft nature upon striking anything solid such as a door, or brick. I say a round with the same dimensions would explode on contact with something solid (again like a brick) as the work-heating would vaporize a piece of aluminum with that mass. If it could even get there without being melted by atmospheric friction. As I've submitted questions to you before we decided we would—drum roll—ASK THE PHYSICIST!!! Are either of us correct? The loser has to buy the next Pizza we order.

Answer: Neither of you is addressing the real issue here—a 4 gram bullet traveling at 80% the speed of light has an almost incomprehensible amount of energy. So, saying that this weapon would 'be useless' is way off the mark. Saying that it is impossible to make such a weapon is a different issue which I will address after talking about what such a speedy bullet would do. You are probably not interested in the details, so I will just give you the kinetic energy such a bullet would have—about 5×10^{14} J. To put that in perspective, if you took that energy and delivered it to the power grid over a period of a day, this would be the equivalent of a 6 gigawatt $(6 \times 10^9 \, \text{W} = 6 \text{ billion J s}^{-1})$ power station. The largest power station in the US has a power output of about 4 gigawatts. Or, the energy of the Nagasaki atomic bomb was about 10^{14} J, 1/5 of the energy of your 4 gram bullet. So to argue whether the bullet's flattening or exploding is the reason it would not do much damage sort of misses the point, don't you think?! If the bullet takes 10 s to deliver its energy to whatever can take it, you are still talking about the energy of 5 WWII-era atomic bombs being delivered. I would not want to be within 50 miles of that. Finally, it means that the 'gun' has to deliver all that energy in an unimaginably short amount of time (I presume that since, if it is 'man-portable', the gun must be no more than a few feet long). I figure that a force on the bullet of more than 2 million lbs would be required during the time it was in the gun; do you think an aluminum bullet could withstand such a force? And, Newton's third law says that, if the gun exerts that force on the bullet, the bullet would exert that force on the gun. Imagine the recoil! This is such a ridiculous scenario, I don't think we even need to beat a dead horse and talk about what air friction would cause to happen as it flew to the target.

Question: What would the yield of a 5000 ton iron slug accelerated at 95% of c by say a bored Omnipotent be? Would it be enough to mass scatter a planet?
Answer: I get the strangest questions sometimes! So, 5000 metric tons $= 5 \times 10^6$ kg. The kinetic energy would be $K = E - mc^2 = mc^2[(1/\sqrt{(1 - 0.95^2)}) - 1] \approx 10^{24}$ J. The energy U required to totally disassemble a uniform sphere of mass M of radius R is $U = 3GM^2/(5R)$ where $G = 6.67 \times 10^{-11}$ is the universal gravitational constant. So, taking the Earth as a 'typical' planet, $U = 3 \times 6.67 \times 10^{-11} \times (6 \times 10^{24})^2/(5 \times 6.4 \times 10^6) \approx 2 \times 10^{32}$ J. So your god's slug is far short of supplying enough energy to totally blast apart the Earth.

Question: I was watching *Babylon 5* and in there was a description of an Earth Alliance space ship weapon. The weapon was a gun that has two very conductive parts on both sides and a conductive armature in the middle and electricity somehow launches the projectile. The gun is 60 m long, has two barrels with each capable of firing two shots per second simultaneously. The projectile is 930 kg in mass (1.7 m long 20 cm in diameter) to a velocity of 41.5 km s^{-1}. The barrel of this electric gun is 60 m long. Is this kind of gun physically possible to build?
Answer: I always like to look first at the energetics when answering questions like this. Assuming that the acceleration of the projectile is uniform, I find that the time it would take to traverse the barrel is 0.029 s and the average acceleration is

1.4×10^6 m s^{-2}. Thus, the average force on the projectile would be $F = ma = 1.4 \times 10^6 \times 930 = 1.3 \times 10^9$ N $= 280\,000\,000$ lb. I do not think you could have a projectile which would not be destroyed by such a force. But, suppose the projectile could withstand this force; the energy which you would have to give it would be $E = \frac{1}{2}mv^2 = 8 \times 10^{11}$ J. Delivering this energy in 0.029 s would require a power input of $P = 8 \times 10^{11}/0.029 = 2.8 \times 10^{13}$ W $= 28$ TW; for comparison, the current total power output for the entire Earth is about 15 TW. Or, if you think of the energy being stored between shots, and there are four shots per second, $P = 8 \times 10^{11}/0.25 = 3.2 \times 10^{12} = 3200$ GW; the largest power plant currently on Earth is about 6 GW. And, this power source needs to be on a ship? I do not think this gun is very practicable!

Question: Hi! I was playing a video game called *Mass Effect 2* and in the game a drill instructor gives a lesson about the main gun of a Mount Everest class dreadnought that is the first dreadnought class made by the largest human government in ME Universe known as Systems Alliance. The main gun is a rail gun that accelerates one 20 kg ferrous slug to 4025 km s^{-1} and it takes 5 s to charge. The slug impacts with kinetic energy of 1.62×10^{14} J which is 38.7 kt of TNT. The main gun is 800 m long and built into the superstructure of an 888 m long dreadnought that has mass of at least one million tons. Would there be any problems of having our future warships operating in space equipped with this kind of weapon and does this design sound feasible? And in the game's lore it is stated that an impact from this weapon levels entire city blocks. Would this projectile moving at 4025 km\s be able to level a city block in a metropolitan area because the Turians (a species in *Mass Effect*) fired these slugs to human cities on the colony of Shanxi leveling a city block clean of even the sturdiest skyscrapers.

Answer: Gamers and sci-fiers have asked questions like this before. That is good, to think about the physics and how it might affect the practicability of these kinds of weapons in the real world. Your numbers are right, the kinetic energy of a 20 kg mass traveling at a speed around 4×10^6 m s^{-1} is about 1.6×10^{14} J. (The speed is just a little above 1% the speed of light, so classical calculations should be ok.) The energy of the atomic bomb dropped on Nagasaki was about 10^{14} J, so this should answer your question about whether or not there is adequate energy to demolish a city block—easily!

- It takes 5 s to charge, so let's see what the required power input would be: $P = E/t = 1.6 \times 10^{14}/5 \approx 3 \times 10^{13}$ W $= 30$ TW. The average power consumption of the entire Earth is 15 TW. Where are you going to get this kind of power in the middle of empty space? Maybe you can just carry hundreds of atomic bombs with you?
- And, let's talk about the launch. If the acceleration over the 800 m is uniform, it would take about 4×10^{-4} s resulting in an acceleration of 10^{10} m s^{-2}. That means that the force necessary to give this acceleration to a 20 kg mass is 2×10^{11} N ≈ 45 billion lb. Do you think an iron slug could withstand such a force?
- Given the time of acceleration, what is the power delivered to the slug? $P = E/t = 1.6 \times 10^{14}/4 \times 10^{-4}$ s $\approx 4 \times 10^{17}$ W.

- Unlike the previous two answers, recoil for this gun should not be a serious problem. If the mass of the ship is a million metric tons, 10^9 kg, the recoil velocity should only be about $20 \times 4 \times 10^6/10^9 = 8$ cm s^{-1}.

I think you would agree that this device would be totally unworkable in the real world.

Added comment: In my comment above stating that recoil would not be a problem, I have to take that back. Even though 8 cm s^{-1} is not very fast, that speed is acquired in a very short time, 4×10^{-4} s, so the acceleration of the ship is very large, $a = 0.08/4 \times 10^{-4} = 200$ m s^{-2} which is approximately $20g$, quite a jolt!

Question: In the *Halo* video game series there are Magnetic Accelerator Cannons in orbit around planets that can launch a 3000 ton magnetic projectile to 4% light speed. These cannons use the principle of the coil gun. These projectiles have a kinetic energy of 2.16×10^{20} J which translates to around 51.6 gigatons of TNT. So these cannons seem to have really unrealistic velocities for these projectiles. What would be the problems in developing these cannons to defend the human species from possible alien invaders? I know energy is one but I've heard that if you were to accelerate a projectile to these kinds of speeds they would turn into plasma from the sheer amount of energy being transferred into them.

Answer: I think you will get the picture of why this is a preposterously impossible weapon if you read an earlier answer. There the speed was much higher but the mass much smaller. Here are the practical problems in a nutshell.

- To accelerate it to this speed in a reasonable distance the force required would be so large as to totally disintegrate the projectile and the cannon for that matter.
- Think about the recoil of the cannon. Unless its mass was much bigger than 3000 tons, much of the energy expended would be wasted, not to mention the disruption of the orbit. This would be a good reason to have it mounted on the ground rather than orbit.
- Where are you going to get the necessary energy? I agree with your number for the kinetic energy of the projectile ($\frac{1}{2}mv^2$ works fine for this relatively low speed and a ton here is a metric ton), ~2.16×10^{20} J. Suppose it took 1 min to get this much energy; then the power required would be $2.16 \times 10^{20}/60 \approx 3.6 \times 10^{18}$ W $= 3.6 \times 10^9$ GW. This is about 1 440 000 times greater than the current total power generated on Earth of about 2500 GW. (Of course, that does not take into account the recoil energy of the cannon itself.)
- Oh yeah, I almost forgot. There is no evidence whatever for alien bad guys.

Finally, one of my favorite books as a kid was *Ringworld* by Larry Niven. Here I calculate the kinematics of projectile motion on such a rotating environment.

Question: Could centrifugal force actually be used to simulate gravity like in so many sci-fi stories? One of my favorite sci-fi stories is the *Ringworld* by Larry Niven. The Ringworld of the title is a giant ring shaped structure the size of

Earth's orbit. It's centered on a star and has a habitable inside edge, gravity on this inside edge is simulated by the structure spinning fast enough to make objects feel as heavy as they would at 99.2% Earth gravity. If I were stood on a real structure like the Ringworld, and I jumped up in the air, would I fall back down or fly off into space?

Answer: Yes, centrifugal force could be used to simulate gravity, as long as the radius of the ring is large compared to the size of the objects. In such a scenario, the centripetal acceleration should be set equal to g, so $g = R\omega^2$ where ω is the angular velocity in radians per second. So, the figure 4.1 shows the ring as viewed from outside. You now jump straight up with a speed v. However, note that you also have a tangential velocity of $R\omega$ so your actual velocity is $\sqrt{(v^2 + R^2\omega^2)}$. So you see what will happen is that you will go in a straight line with constant speed (because there are no forces on you) along that velocity until you again strike the ring; it will seem that you jumped and came back down. You can calculate the time you were in the air and how high you went by doing some pretty straight-forward geometry/trigonometry.

I find the two angles labeled θ above are the same so

$$\sin\theta = v/\sqrt{(v^2 + R^2\omega^2)} \qquad \text{and} \qquad \cos\theta = R\omega/\sqrt{(v^2 + R^2\omega^2)}.$$

From these you can find the length of the chord (the length of your flight)

$$C = 2R\sin\theta = 2Rv/\sqrt{(v^2 + R^2\omega^2)},$$

the height h you go above the surface

$$h = R(1 - \cos\theta) = R\left(1 - R\omega/\sqrt{(v^2 + R^2\omega^2)}\right),$$

and the time you are in the air

$$T = C/\sqrt{(v^2 + R^2\omega^2)} = 2Rv/(v^2 + R^2\omega^2).$$

Now, let's compare these with the time and height on Earth. Remember that $\omega = \sqrt{(g/R)}$ and I will take R to be very large compared to v^2/g. So, now I find that

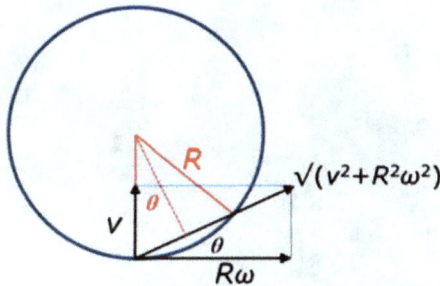

Figure 4.1. The path of a 'vertically' launched projectile in a rotating gravity simulator.

$$h = R\left(1 - R\omega/\sqrt{\left(v^2 + R^2\omega^2\right)}\right)$$
$$= R\left(1 - 1/\sqrt{\left(1 + v^2/\left(R^2\omega^2\right)\right)}\right)$$
$$= R\left[1 - \left(1 + v^2/(Rg)\right)^{-\frac{1}{2}}\right]$$
$$\approx R\left[1 - \left(1 - \tfrac{1}{2}v^2/(Rg) + \cdots\right)\right]$$
$$\approx v^2/(2g)$$

and

$$T = 2Rv/\left(v^2 + R^2\omega^2\right)$$
$$= 2R/\left(v\left(1 + R^2\omega^2/v^2\right)\right)$$
$$= 2R/\left(v\left(1 + Rg/v^2\right)\right)$$
$$\approx 2v/g.$$

These two approximate results are just the same as for a projectile launched straight up on Earth. Keep in mind, though, that R must be large. Also, you do not want to jump with a velocity which has a component parallel to the axis of rotation. If you are too close to the edge, you will miss the ring when you 'fall back'. (See appendix D.)

Chapter 5

Epilogue

5.1 Ask the psychic

It is amazing how many visitors to the *Ask the Physicist* web site think they are asking the psychic. In spite of the disclaimers I have put on the home page, nearly every day I receive at least one question asking me to use my mystical powers to help guide their daily lives or explain remarkable phenomena. And, it is not just ignorant people who cannot spell—computer algorithms don't seem to get it either. A few years ago I tried to use GoogleAds on the site to generate a little money; to my horror, many ads generated by Google's software were for 'Free Psychic Readings'! And, if you have ever tried to talk to a human being at Google, forget it; I just gave up using GoogleAds when I couldn't stop the psychic ads—what better way to destroy the credibility of a serious science site than with ads for palmists and clairvoyants? Gathered here are some of my favorites from Ask the Psychic. (By the way, unlike the serious parts of this book, questions in this chapter are not edited.)

> **Question:** Me and my gf were chatting one night over the computer. I was thinking about holding her hand and she immediately tells me her hand got warmer we kept experimenting with this until I could see every detail about her room even though I've never seen her room. Eventually after about 3 h of experimenting I started to read her thoughts. Even stranger still I asked her to put her hand on her computer and when I did I felt myself get shocked with like an other worldly energy and she felt it too. Now I know about half her memories and she knows mine. Help! I don't know what's going on.
>
> **Question:** Will Cynthia come back and want to be with me?
>
> **Question:** Hello, my name is John and I have just one question. If I try to communicate with my loved ones that have passed, is it possible to help me in life?
>
> **Question:** Where did my sons hampster disappears to?
>
> **Question:** Okay so I REALLY like this kid Branden. And I'm having dreams about me and him dating. I talk to him everyday we text, video chat and call each other. I have recently moved so we stay in touch. He's told me that he thinks I'm

5-1

cute and pretty. But he likes someone other than me. So you knowing that, do you think me and him will EVER get together?

Question: where is my wii controller

Question: Have Jermaine Hogan ever cheated on me

Question: What did I see? I few days ago I was sitting in front of my pc reading something. I took a casual glance down my body (chest) and saw kind of vapor coming out from my body, surrounding it. Later on I found On YouTube a film showing humans radiating heat captured by an infrared cam. What I saw looked a bit like that. The movement of this 'fog' or 'vapor' reminded me of real vapor.

Question: I would liked to know about my future and do I actually have a twin brother who know I'm still here waiting for him.

And, finally, the perennial favorite,

Question: am I pregnant.

5.2 Off-the-wall hall of fame

Although I emphasize that I require single, concise, well-focused questions, inevitably I get lengthy questions (dissertations, really) from people wishing to get the stamp of approval of a professional physicist on their personal theory of something or other. If these submissions are not questions but rather assertions, then they are often questions which start with something like "…is it possible that…" Of course a scientist is often loathe to say that something is impossible even if he believes it to be—then you get attacked for being closed-minded. Best to just file these away in my Hall of Fame rather than try to get into a dialog with a crazy person! Here are a few good ones.

Question: in fullmetal alchemist alchemy is a scinace of understading mater then bracking it down and recontucting it why cant a black hole do that just in a difrent part of the universe. i firlmy beliove that black holes are cheat cuts to difrent parts of space. i just dont have the brain to prove it in a sctaic eqation but i qam will to work with those who have the resorces and mind to do so. i would like to be apart of a group to do this.

Question: I was wondering about the propagation of our AC/DC electricity and I was blessed with the thoughts of an electrostatic vortex that could be created by properly encasing the wires and putting them closer to the ground, if not underground, to allow the power a conduit that does not have to work against gravity any harder. I considered what is the best conduit for static electricity and I thought wool! We can create a fluffed wool mesh that can encase power lines and that will be encased with a mirrored lining and non conducive rubberized plastic casings to reflect free falling electrons (power loss) that will magnetically charge the wool as they pass back and forth through it until they begin to cling to the mesh and create an electromagnetic vortex surrounding the whole of the current propagated so that there will be no loss and discharge of electrons and power can begin to flow without power loss. It would be simple to configure and design an electrostatic alternator and diodes to change electromagnetic static electricity back into AC/DC current. The current that comes through the lines to household will be an even more pure sine wave and it will get the atrocious power lines, and buzz, out of the air. What do you think? Do you understand what I mean? I think it was where Tesla was headed but he thought just to remove static electricity from the air. The fact that we release our static electricity into the air, creating a surplus of it, may be detrimental to environmental health and it may also affect the orbit of the earth. Statically charged radio frequencies can create electromagnetic vortexes in the atmosphere that can affect synaptic flow and electrical equipment failure and malfunction. We need to secure this invisible enemy and begin to get our moneys' worth from it.

Question: I have some questions which are really important to me because they have literally speaking ruined my life and have caused a lot of phobias in my mind and soul, sir, i have a strange feeling everytime that anytime the earth will start rotating and i would not be able to sustain that and will die due to my motion phobia. Sir, is it really possible that the earth will start rotating and we will be able to see it? (I do know that earth is rotating with a constant acceleration) but could anything happen that would disrupt the constant nature, and we will be able to see it? i mean as a lot of traffic is moving on the surface, so they should cause acceleration and deacceleration to the earth's movement. sir, plz reply me asap as i can't travel by any means due to this circular movent phobia and everything.

Question: this question is concerning negative matter (−e) and positive matter (+e) in relation to wormholes

1 is the opposite of −1.

1 is 'more than everything'

as we know 1 is more than nothing

1 is also less than 2

therefore 2 is more than more than everything or 1 is less than everything

1 is more than nothing −1 is less than everything

$1 = −1$ 1 and 1 are opposites 1,1,−1 1>−1

$−1 = 1$ 1 and −1 are opposites of

is 1 is a positive matter (+e) than that positive and two opposites, the opposite of a positive is a negative 1 and −1 are negative, but 1 = 1 so for every +e there is a −e which is of equal value and a −e which is of lesser value

I am lost in all of this and how it could be applied to negative matter in black holes and their wormholes(if present)?

Question: A step beyond the physics of the physical world there is the precept of meta-physics or that spirit known as 'Karma' in the world. The precepts of ethical force that becomes contrived by the actions of one that bring either falter or reward from that spirit of ethics and the force that our higher power has invested in its effects as per action/reaction. In this sense, would it stand to reason that in the use of trees for paper and paper products (so frivolous a use of such a precious resource) when there is highly renewable water grasses (cypress and papyrus grasses) and modern technology to alleviate such a waste of our precious trees. In that we have not developed a process to derive paper from these alternate resources I believe that the human condition suffers from the spirits of the forests that are as alive as any self respecting human being. I also believe that the animal populations of the water punish us (degrade the quality of human life) for not cleaning our waste waters more effectively by recreating the natural osmosis that promotes the sedimentary processing of the waters in underground systems. We could recreate the conditions of underground lakes and create an electromagnetic osmosis that we could intensify to make a sedimentary time machine and do in 1 year what takes 1000 naturally. Do not even answer this question because I know I am right; just use sense and get to work on these projects. I have many more to follow.

From Newton to Einstein
Ask the physicist about mechanics and relativity
F Todd Baker

Appendix A

Energy

A1.1 Work-energy theorem in Newtonian mechanics

Although I use calculus so that I have generalized to forces which might vary, I do the calculations in one dimension for clarity. First, derive the work–energy theorem in Newtonian physics:

$$F = \mathrm{d}p/\mathrm{d}t$$
$$= m(\mathrm{d}v/\mathrm{d}t)$$
$$= m(\mathrm{d}v/\mathrm{d}x)(\mathrm{d}x/\mathrm{d}t)$$
$$= m(\mathrm{d}v/\mathrm{d}x)v.$$

Rearranging,

$$F\mathrm{d}x = mv\mathrm{d}v.$$

Integrating,

$$W = \int F\mathrm{d}x = m\int v\mathrm{d}v = \frac{1}{2}mv_2^2 - \frac{1}{2}mv_1^2 = \Delta K.$$

A1.2 Potential energy

Suppose there is some force, call it the *internal* force F_{int}, which is always present for some particular problem and that all other forces doing work are represented by F. Then

$$W = \int F\mathrm{d}x + \int F_{\mathrm{int}}\mathrm{d}x = W_{\mathrm{ext}} + \int F_{\mathrm{int}}\mathrm{d}x = \Delta K$$
$$\Delta K - \int F_{\mathrm{int}}\mathrm{d}x = W_{\mathrm{ext}}.$$

doi:10.1088/978-1-6270-5497-3ch6

Now, the indefinite integral $-\int F_1 dx$ is just some function of x, call it $U(x)$, so

$$-\int F_1 dx = U(x_2) - U(x_1) = \Delta U(x).$$

U is called the potential energy. So now we define the total energy as $E = K + U$ and rewrite the work–energy theorem as

$$\Delta E = W_{ext}$$

where W_{ext} is the work done by external forces which are all forces for which a potential energy function has not been included in the energy. So, you see, the potential energy is really just a clever book-keeping device to keep track of work done by a force which is always present.

In this book, you will seldom need to understand potential energy. The one simple case which you might need is the gravitational potential energy. In that case the force is $F_1 = -mg$ (the $-$ sign because it points down) and so $U_{grav}(y) = mgy$. (We usually use y for the vertical direction rather than x.)

A1.3 Energy in special relativity

In the theory of special relativity everything is the same except p is redefined:

$$F = dp/dt$$

$$= m\left(d\left[v/\sqrt{(1 - v^2/c^2)}\right]/dt\right)$$

$$= m\left(d\left[v/\sqrt{(1 - v^2/c^2)}\right]/dx\right)(dx/dt)$$

$$= mv\left(d\left[v/\sqrt{(1 - v^2/c^2)}\right]/dx\right)$$

$$= mv\left(1 - v^2/c^2\right)^{-3/2}dv/dx.$$

Rearranging,

$$F dx = mv\left(1 - v^2/c^2\right)^{-3/2}dv.$$

Integrating,

$$W = \int F dx = m\int v\left(1 - v^2/c^2\right)^{-3/2}dv = mc^2(\gamma_2 - \gamma_1) = \Delta K.$$

If the particle started from rest, $\gamma_1 = 1$ and ended at speed v, $\gamma_2 = [(\gamma - 1)/\sqrt{(1 - v^2/c^2)}]$, then

$$K = mc^2(\gamma - 1).$$

Be sure to note that m is the rest mass. Now, this does not look much like $\frac{1}{2}mv^2$ for small v, so we need to look a little more closely. If $v \ll c$,

$$1/\sqrt{(1 - v^2/c^2)} \approx 1 + \frac{1}{2}v^2/c^2 + \cdots$$

This is just a binomial expansion, $(1 + z)^n \approx 1 + nz + 1/2n(n - 1)z^2 + \cdots$ So now we can write

$$K \approx mc^2\left(1 + \frac{1}{2}v^2/c^2 + \cdots - 1\right) \approx \frac{1}{2}mv^2.$$

Rearranging the equation for K above,

$$K = \gamma mc^2 - mc^2.$$

This equation says that 'the kinetic energy is something minus some constant'. We interpret this to mean that the 'something' is the total energy E and the constant mc^2 is the energy something has by virtue of its mass, even if at rest. So, we can finally write the total energy as $E = \gamma mc^2$. So, if the particle is at rest, $\gamma = 1$ and $E = mc^2$. Although I will not work it out, a little algebra leads to the very useful expression for the total energy in terms of the momentum:

$$E = \gamma mc^2 = \sqrt{(p^2c^2 + m^2c^4)}.$$

Appendix B

Approximations in Kepler's laws

The purpose of this appendix is to show that the last small distance of the Kepler straight-line orbit which I did in an example in chapter 1 does not significantly contribute to the total time; also to show that the speed does not become relativistic until very small distances from the force center. The radius of the Earth's orbit is $R_O = 1.5 \times 10^{11}$ m and the radius of the Sun is $R_S = 7 \times 10^8$ m. The mass of the Earth is $M_E = 6 \times 10^{24}$ kg and the mass of the Sun is $M_S = 2 \times 10^{30}$ kg. The potential energy function is $U(r) = -GM_S M_E/r$. So, for the Earth stopped in its orbit its total energy is

$$E_1 = -GM_S M_E/R_O$$

and the energy at the Sun's surface, now with velocity V, is

$$E_2 = -GM_S M_E/R_S + \frac{1}{2}M_E V^2.$$

Conserving energy, $E_1 = E_2$ and solving for V,

$$V = \sqrt{\{2GM_S[(1/R_S) - (1/R_O)]\}}$$
$$= 1.63 \times 10^{10}\sqrt{[(1/7 \times 10^8) - (1/1.5 \times 10^{11})]}$$
$$= 1.25 \times 10^6 \text{ m s}^{-1}.$$

The first thing to note is that this speed is still much smaller than the speed of light, about $0.004c$, so relativistic corrections are not necessary. If the Sun and Earth were point masses, the Earth would continue accelerating from here. If it kept going at speed V it would take a time $t = R_S/V = 560$ s to reach the center; the actual time would be much shorter than this because of the acceleration. But the total time to fall to the surface was found to be about 65 days, so we can conclude that the error made in the approximations was truly negligible.

doi:10.1088/978-1-6270-5497-3ch7

From Newton to Einstein
Ask the physicist about mechanics and relativity
F Todd Baker

Appendix C

Rotational physics

Newton's laws, as described in chapter 1, section 1.1, applies to objects which can do nothing but move along lines in response to forces; this kind of motion is called translational motion. But, this is not the most general way objects can move. Imagine a stick which you have grasped at one end and thrown; the stick does indeed seem to follow a path like a small ball would, but it also spins like a propeller as it moves along that path. Objects can also rotate and Newton's laws need to be extended to include the possibility of rotation. Description of rotational physics often runs to two to three chapters in a physics textbook and most of the formalism is not needed here. It is, though, important to the purposes of this book to include some basic ideas.

Torque plays the role of force. Where a net force causes an acceleration (speeding up or slowing down), a net torque causes an angular acceleration (spinning faster or slower). Torque is a little trickier than force to understand because it matters where the force is applied and in what direction. My favorite way to explain it is to imagine closing a door. You must push on the door somehow. But suppose you push on the edge of the door where the hinges are—it will not close no matter how hard you push. Now suppose you push on the edge of the door opposite where the hinges are but you push straight toward the hinges—it will not close no matter how hard you push. Torque τ is defined as the component of the force perpendicular $F_\perp$ to the moment arm L times the length of the moment arm, $\tau = F_\perp L = FL \sin \theta$. You must choose an axis around which to calculate torques. If you call the torque positive as I have for the example in figure C.1, and it tends to cause the moment

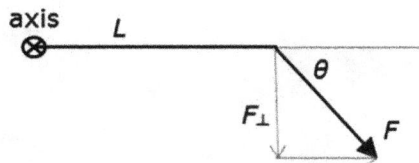

Figure C.1. A force F exerting a torque.

doi:10.1088/978-1-6270-5497-3ch8   C-1

arm to rotate clockwise as you can see it does, then every other torque which tends to cause clockwise rotation must also be positive and torques which tend to cause counterclockwise rotation must be negative. Most of the questions answered are equilibrium problems where all the torques add to zero. If there is a net torque, there will be an angular acceleration α and Newton's second law takes the form $\tau = I\alpha$ where I is the moment of inertia, that which plays the role of mass in rotational physics. You need not worry about how to calculate I, it will always be given in any problems in this book. For example, the moment of inertia of a solid uniform disk of radius R and mass M is $I = 1/2MR^2$. The name makes sense because if m is the resistance to acceleration (inertia), I is the resistance to angular acceleration (moment of inertia).

Appendix D

Centripetal acceleration

Acceleration is a vector, it has both a direction and a magnitude. So, contrary to common nomenclature, a car driving around a curve with constant speed is accelerating. And, you should know that you cannot turn a corner going a constant speed on an icy road, that is, without the force provided by the friction between your tires and the road. So, if you believe Newton's second law, there must be some acceleration even if you are going with a constant speed. You usually think of acceleration as change in speed, but when you are turning there is also an acceleration due to the change in direction of your velocity. This acceleration is called centripetal acceleration. This word, *centripetal*, has its origin in Latin: the *centri-* part is pretty obvious, center, the *-petal* part is from the Latin *peto*, I seek; centripetal acceleration points toward the center of the circle you are traveling in. You may know the word *centrifugal*. Here, *-fugal* comes from the Latin *fugo*, I flee. Centrifugal acceleration points away from the center of the circle you are traveling in and there is no such thing although we will find the (fictitious) concept of centrifugal acceleration useful. Think of a stone tied to a string of length R and moving with some constant speed v. Would the stone move in a circle if the string were not there? Of course not. The string exerts a force (called the tension in the string) on the stone which points along the string at the center of the circle. Any introductory physics text can give you a derivation of the magnitude of the centripetal acceleration a_c, I just give the result, $a_c = v^2/R$.

doi:10.1088/978-1-6270-5497-3ch9

From Newton to Einstein
Ask the physicist about mechanics and relativity
F Todd Baker

Appendix E

Friction

When two surfaces are in contact with each other and sliding, they exert forces on each other. The force which is parallel to the surface of contact is called the frictional force. Friction can be exceedingly complicated, but for many real-world situations it is true that the frictional force f is approximately proportional to how hard the two surfaces are pressed together; that force is usually called the normal force N and is the force which the two surfaces exert on each other perpendicular to their surfaces. So, $f \propto N$. The simplest example is if the surfaces are horizontal so that $N = mg$ (because of Newton's first law). For any given situation you need to measure both f and N to find the proportionality constant to make this an equation, $f = \mu_k N$. μ_k is dimensionless (a ratio of forces) and called the coefficient of kinetic friction and, to a very good approximation for every-day situations, depends only on the materials in contact; μ_k is large for rubber on dry asphalt (0.5–0.8) and small for teflon on teflon (0.04), for example.

   If the surfaces are not sliding, they may or may not exert forces parallel to the surfaces on each other. For example, if a book sits on a horizontal table, there is no frictional force. But, if you push gently horizontally, it does not move so there must be a frictional force equal in magnitude but in the opposite direction as your force. As you push harder and harder, the friction gets bigger and bigger until, eventually, the book pops away. The maximum frictional force you can get, f_{max}, is proportional again to N and the proportionality constant is the coefficient of static friction, μ_s; $f_{max} = \mu_s N$. In general, the static frictional force may be written as $f < \mu_s N$. It is always true that $\mu_s > \mu_k$ because when f_{max} is reached the formerly static object accelerates.

doi:10.1088/978-1-6270-5497-3ch10

From Newton to Einstein
Ask the physicist about mechanics and relativity
F Todd Baker

Appendix F

The constants of electricity and magnetism

Most textbooks introduce the unit of charge (coulomb) before the unit of current (ampere). I have done it a little differently here because the thing which is actually operationally defined is the ampere. Since this book does almost no electricity and magnetism, there is no real reason to introduce the coulomb first. In addition, most laypersons have a much better idea what an ampere is than what a coulomb is.

Two long parallel wires, each of length L and separated by a distance r carry electric currents I_1 and I_2. They are observed to exert equal and opposite forces (Newton's third law) on each other and the magnitude of this force is proportional to each of the currents, the length of the wires, and inversely proportional to the separation: $F \propto LI_1I_2/r$. Choosing a proportionality constant $\mu_0/(4\pi) = 10^{-7}$ defines what the unit of electric current is: $F/L = \mu_0 I_1 I_2/(2\pi r)$. So, if two wires are carrying equal currents, are separated by 1 m, and the force per meter each wire experiences is $2 \times 10^{-7}\,\mathrm{N\,m^{-1}}$, then each wire is carrying an electric current of 1 ampere (A). Since electric current is the rate at which electric charge is flowing, knowing the ampere also lets us know the electric charge unit we will use, called the coulomb (C) because $1\,\mathrm{A} = 1\,\mathrm{C\,s^{-1}}$. To set the scale relative to everyday life, 1 A is a typical household current. The charge on an electron is $-1.6 \times 10^{-19}\,\mathrm{C}$, so a current of 1 A corresponds to $1/1.6 \times 10^{-19} = 6.25 \times 10^{18}$ electrons $\mathrm{s^{-1}}$. So, the first constant μ_0, called the permeability of free space and which sets the scale of magnetic fields in the system of units we use, is exactly (because we defined it that way)

$$\mu_0 = 4\pi \times 10^{-7}\,\mathrm{N\,A^{-2}} = 4\pi \times 10^{-7}\,\mathrm{N} \times \mathrm{s^2\,C^{-2}}.$$

We now know what a coulomb is. If we go to a laboratory and measure the force F between two electric charges, Q_1 and Q_2, separated by some distance r we find that $F \propto Q_1Q_2/r^2$. Now, to make this an equation we need to measure the proportionality constant because we know how charge and length are measured. Doing this, we find that $F = Q_1Q_2/(4\pi\varepsilon_0 r^2)$; this is called Coulomb's law. Note that we have chosen to write the proportionality constant (which we have measured) as $1/(4\pi\varepsilon_0)$. So, the second constant ε_0, called the permittivity of free space and which sets the scale of electric fields in the system of units we use, is exactly (because we measured it)

$$\varepsilon_0 = 8.85 \times 10^{-12} \ \mathrm{C^2/(N{\cdot}m^2)}.$$

Maxwell's equations predict waves which have a velocity of $1/\sqrt{(\varepsilon_0\mu_0)} = 3 \times 10^8 \ \mathrm{m \ s^{-1}}$. This is truly one of mankind's most remarkable intellectual achievements!

Appendix G

Galilean and Lorentz transformations

If one inertial frame (x, y, z, t) is at rest and another (x', y', z', t') moves in the $+x$ direction with speed v, and at $t = t' = 0$ the origins were coincident, the equations of Galilean relativity are:

$$x' = x - vt$$
$$y' = y$$
$$z' = z$$
$$t' = t$$

Figure G.1 shows a blue ball with its x-coordinates indicated at time $t = t'$. The ball also has velocity u and acceleration a, both in the x-direction. To find u' and a' is simply a matter of applying the definitions of velocity and acceleration:

$$u' = dx'/dt' = d/dt\,(x - vt) = dx/dt - v = u - v$$
$$a' = du'/dt' = du'/dt = du/dt - dv/dt = a - 0 = a.$$

Note that the assumption most wrong is that clocks in both coordinate systems run at the same rate.

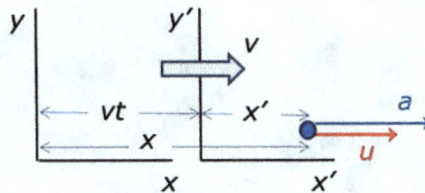

Figure G.1. Two inertial frames with relative velocity v.

doi:10.1088/978-1-6270-5497-3ch12

The corresponding Lorentz transformations are:

$x' = \gamma(x - vt)$

$y' = y$

$z' = z$

$t' = \gamma(t - vx/c^2)$

$u' = (u - v)/(1 + uv/c^2)$

a' is too complicated to write and of no particular use.

www.ingramcontent.com/pod-product-compliance
Lightning Source LLC
Chambersburg PA
CBHW071940220326
41599CB00033BA/6630